FUTURE PERFECT?

FUTURE PERFECT?

God, Medicine and Human Identity

Edited by
Celia Deane-Drummond
and
Peter Manley Scott

t & t clark

Published by T&T Clark
A Continuum imprint

The Tower Building, 11 York Road, London SE1 7NX
80 Maiden Lane, Suite 704, New York, NY 10038

www.tandtclark.com

First published 2006. Paperback edition published 2010.

British Library Cataloguing-in-Publication Data
A catalogue record for this book is available from the British Library

Typeset by Fakenham Photosetting Limited, Fakenham, Norfolk

ISBN 978-0-5670-3079-5 (hardback)
ISBN 978-0-5672-3401-8 (paperback)

CONTENTS

Acknowledgements

The idea for this book first emerged in a conversation between Celia Deane-Drummond and Peter Francis, warden of St Deiniol's library, in March 2004. We discussed the possibility of hosting at the library an expert colloquium relating medical, theological and philosophical issues in the summer of 2005. Further conversations between Celia Deane-Drummond and Fiona Murphy of Continuum in April 2004 at the European Society for the Study of Science and Theology meeting in Barcelona finalized the idea. After some conversations with the Dean of Chester Cathedral, the Very Rev. Professor Gordon McPhate, who was formerly a senior lecturer in pathology at the University of St Andrews, Celia Deane-Drummond invited a number of colleagues to contribute to the colloquium and to a volume of essays arising out of that colloquium. We were also keenly aware that it would be very helpful to invite a diverse group of participants who had a particular interest in the field and were able to offer critical feedback and suggestions as to how the book could develop. The enthusiasm with which this project was greeted in many quarters was very gratifying; Continuum/T & T Clark International agreed to publish the papers from the colloquium.

The following were participants at the colloquium and generously gave of their time and expertise: Dr Peter Manley Scott, Dr David Jones, Dr Julie Clague, Dr Robert Song, Dr Bronislaw Szerszynski, Dr Lisa Goddard, Rev. Mark Bratton, Rev. Stephen Bellamy, Prof. Ulf Görman, Prof. Maureen Junker-Kenny, Dr Michael Northcott, The Very Rev. Prof. Gordon McPhate, Prof. Gordon Graham, Prof. Gareth Jones, Dr Neil Messer, Prof. Celia Deane-Drummond, Dr Brent Waters and Anne Marie Sowerbutts. Administrative aspects of the colloquium were dealt with by the hard work of Anne Marie Sowerbutts, research assistant to the Centre for Religion and the Biosciences, with some able help from Jackie Turvey, a doctoral student at the University of Chester. Peter Manley Scott joined Celia Deane-Drummond as co-editor following the colloquium.

Two of the chapters in this book are revised versions of journal articles. Jenny Kitzinger's and Clare Williams' chapter has been published in a different version as J. Kitzinger and C. Williams, 'Forecasting Science Futures: Legitimising Hope and Calming Fears in the Embryo Stem Cell Debate', *Social Science and Medicine* 61 (2005), pp. 731–40, Copyright 2005. Elaine Graham's chapter will be published in a different version in *Ecotheology* 11.2 (2006) under the title 'In Whose Image? Representations of Technology and the "Ends" of Humanity'. The editors wish to thank the publishers Elsevier and Equinox, respectively, for kindly giving permission for these articles to be reprinted.

This project would have been impossible without the financial assistance from the affiliation of the Centre for Religion and the Biosciences at the University of Chester with Metanexus (Philadephia, USA) as a Local Society Initiative. The project took place in the second year of a three-year programme of funding. St Deiniol's library also generously contributed to the accommodation costs through their scholarship scheme. Celia Deane-Drummond is also grateful to the University of Chester for the provision of a period of sabbatical leave in the academic year of 2004/5, which assisted in the bringing of this project to completion.

Due acknowledgement also needs to be given to Fiona Murphy, Philip Law and latterly Rebecca Vaughan Williams of Continuum for their support and encouragement while this project has been developed.

CONTRIBUTORS

Celia Deane-Drummond has degrees and doctorates in plant science and theology, and her research has focused particularly on the interrelationship between Christian theology and the biological sciences. She received a personal Chair in theology and the biological sciences from the University of Chester in 2000. She has lectured widely both nationally and internationally and has published numerous articles and contributed to books in the field including, more recently, *The Ethics of Nature* (Blackwells, 2004). She is also co-editor with B. Szerszynski of *Reordering Nature: Theology, Society and the New Genetics* (Continuum/T & T Clark, 2003) and editor of *Brave New World: Theology, Ethics and the Human Genome* (Continuum/T & T Clark International, 2003). In 2002 the Centre for Religion and the Biosciences at the University of Chester was formally launched, directed by Celia Deane-Drummond. She was elected Fellow of the Royal Society of Arts and Commerce (FRSA) in 2004.

Ulf Görman is professor of ethics at Lund University, Sweden. Görman is past president of ESSSAT, the European Society for the Study of Science and Theology. His research focuses on bioethics and on ethical issues in science and religion. He is currently heading a national research programme, where Swedish ethicists investigate ethical aspects of postgenomic research. Current publications include articles on bioethics and the anthologies *Etik och genteknik* (in Swedish; Nordic Academic Press, 2004) and *Creative Creatures: Values and Ethical Issues in Theology, Science and Technology* (Continuum/T & T Clark, 2005).

Elaine Graham has worked at the University of Manchester since 1988. She is currently Samuel Ferguson Professor of Social and Pastoral Theology. She is a graduate of the University of Bristol (1980) in Sociology and Economic and Social History. She worked as Northern Regional Secretary for the Student Christian Movement in the early 1980s and as ecumenical lay chaplain at Sheffield City Polytechnic (now Sheffield Hallam University) between 1984 and 1988. Her postgraduate work was completed part-time at the University of Manchester: her MA dissertation (1988) was on the pastoral needs of women, and her PhD thesis (1993) was a critique of theories of gender and their implications for theological debate. Her publications include *Life-Cycles: Women and Pastoral Care* (SPCK, 1993, co-edited with Margaret Halsey), *Making the Difference: Gender, Personhood and Theology* (Mowbray, Fortress, 1995), *Transforming Practice* (Mowbray, 1996), *Representations of the Post/Human* (Manchester University Press, 2002) and numerous articles and reviews. She has just completed a volume entitled *Theological Reflection: Methods* (2005, with Heather Walton and Frances Ward) for SCM Press. She is also a member of the Archbishops' Commission on Urban Life and Faith, due to report in the spring of 2006 to mark the 20th anniversary of *Faith in the City*.

Gordon Graham is Luce Professor of Philosophy and the Arts at Princeton Theological Seminary, New Jersey, and a Fellow of the Royal Society of Edinburgh. He was formerly Regius Professor of Moral Philosophy at the University of Aberdeen in Scotland. He has lectured widely on medical ethics, and his publications include *Evil and Christian Ethics* (Cambridge University Press, 2001), *Genes: A Philosophical Inquiry* (Routledge, 2002) and *Eight Theories of Ethics* (Routledge, 2004).

Søren Holm is a medical doctor and philosopher. He holds a professorial fellowship in bioethics at Cardiff Law School and a part-time chair in medical ethics at the Section for Medical Ethics, University of Oslo. He has a special interest in the specific problems of validity and soundness in bioethical arguments.

D. Gareth Jones is Deputy Vice-Chancellor (Academic and International) and Professor of Anatomy and Structural Biology at the University of Otago, Dunedin, New Zealand. He trained in medicine and neuroscience at University College London and University College Hospital. Recent books include *Valuing People* (Paternoster, 1999), *Speaking for the Dead* (Ashgate, 2000), *Clones* (Paternoster, 2001), *Stem Cell Research and Cloning* (Australian Theological Forum Press, 2004), and *Designers of the Future* (Monarch, 2005). He is co-author of *Medical Ethics* (Oxford University Press, 2005, 4th edn). He was made Companion of the New Zealand Order of Merit (CNZM) in 2004 for his contributions to science and education; he holds the degrees of DSc and MD for his work in neuroscience and bioethics respectively.

Maureen Junker-Kenny is Associate Professor of Theology at Trinity College, Dublin where she teaches Christian Ethics/Practical Theology. She gained a PhD in 1989 at the University of Münster, on F. Schleiermacher's theory of religion and Christology. The theme of her *Habilitationsschrift* in 1996 at the University of Tübingen was J. Habermas's ethics of argumentation. She has published in fundamental theology, foundations of philosophical and theological ethics, and bioethics, editing *Designing Life? Genetics, Procreation, and Ethics* (Ashgate, 1999), and co-editing *The Discourse of Human Dignity* (Concilium 2003/2; SCM Press, 2003), *Memory, Narrativity, Self, and the Challenge to Think God: The Reception within Theology of the Recent Work of Paul Ricoeur* (LIT-Verlag, 2004).

Jenny Kitzinger is Professor of Media and Communication Research at Cardiff University. She is author of *Framing Abuse: Media Influence and Public Understanding of Sexual Violence against Children* (Pluto Press, 2004), and co-author of *The Mass Media and Power in Modern Britain* (Oxford University Press, 1997) and *The Circuit of Mass Communication in the AIDS crisis* (Sage, 1997). She is also co-editor of *Developing Focus Group Research* (Sage, 1999).

Gordon McPhate is a member of the Society of Ordained Scientists and trained consecutively as a physician and an Anglican priest, combining both vocations in joint pastoral and clinical academic posts at the Universities of London (Physiology) and St Andrews (Pathology). Subsequent to a master's degree and a research doctorate, he qualified as an endocrinologist and chemical pathologist, and became a hospital consultant in these fields, specializing in diabetes. He is additionally qualified in the fields of medical ethics and neuroscience, which he has taught to medical students over many years. Currently, he is Visiting Professor of Theology and Medicine at the University of Chester, in addition to his substantive post as Dean of Chester Cathedral.

Neil Messer is Lecturer in Christian Theology and Director of the Research Centre for Contemporary and Pastoral Theology at the University of Wales, Lampeter. After doctoral research in molecular biology, he embarked on theological studies in Cambridge and was ordained as a minister of the United Reformed Church in 1993. Following postgraduate studies in theology, he taught Christian Ethics at Mansfield College, Oxford, and the Queen's Foundation, Birmingham, before moving to Lampeter in 2001. He is the author of *Study Guide to Christian Ethics* (SCM Press, 2006), *The Ethics of Human Cloning* (Grove, 2001) and various articles and reviews on bioethics, and editor of *Theological Issues in Bioethics: An Introduction with Readings* (Darton, Longman & Todd, 2002).

Michael S. Northcott is Reader in Christian Ethics in the School of Divinity at the University of Edinburgh, Associate Rector at St James' Leith in the Scottish Episcopal Church, and Canon Theologian of Liverpool Cathedral. He has written widely on environmental and genetic ethics and is currently working on the ethics of climate change. He was a principal contributor to *Engineering Genesis: The Ethics of Genetic Engineering in Nonhuman Species* (Earthscan, 1998), and he is author of *The Environment and Christian Ethics* (Cambridge University Press, 1999), *An Angel Directs the Storm: Apocalyptic Religion and American Empire* (I.B. Tauris, 2004) and *A Moral Climate? The Ethics of Global Warming* (Darton, Longman & Todd and Brazos, 2006).

Ted Peters is Professor of Systematic Theology at Pacific Lutheran Theological Seminary and the Graduate Theological Union in Berkeley, California. He is a research associate with the Center for Theology and the Natural Sciences. He is editor of *Dialog, A Journal of Theology* and co-editor of *Theology and Science.* He is author of *Science, Theology, and Ethics* (Ashgate, 2003), *Playing God? Genetic Determinism and Human Freedom* (Routledge, 2002) and *GOD – The World's Future: Systematic Theology for a New Era* (Fortress, 2000). He is co-author of *Evolution from Creation to New Creation* (Abingdon, 2003).

Peter Manley Scott is Senior Lecturer in Christian Social Thought and Director of the Lincoln Theological Institute at the University of Manchester. He is author of *Theology, Ideology and Liberation* (Cambridge University Press, 1994), *A Political Theology of Nature* (Cambridge University Press, 2003) and numerous articles, and is co-editor of the *Blackwell Companion to Political Theology* (2004). He is a member of the Center of Theological Inquiry (Princeton, USA) and is on the editorial board of the journal *Ecotheology*. He is currently writing a book on theological anthropology for an ecological, global age.

Brent Waters is Director of the Jerre L. and Mary Joy Stead Center for Ethics and

Values, and Associate Professor of Christian Social Ethics at Garrett-Evangelical Theological Seminary, Evanston, Illinois. He is the author of *From Human to Posthuman: Christian Theology and Technology in a Postmodern World* (Ashgate, 2006), *Reproductive Technology: Towards a Theology of Procreative Stewardship, Dying and Death: A Resource for Christian Reflection* (Pilgrim Press, 2001), and *Pastoral Genetics: Theology and Care at the Beginning of Life* (with co-author Ronald Cole-Turner; Pilgrim Press, 1996), and co-editor of *God and the Embryo: Religious Voices on Stem Cells and Cloning* (with Ronald Cole-Turner; Georgetown, 2003). Waters has also written numerous articles and lectured extensively on the relationship between theology, ethics and technology. He is currently pursuing a research project on Christian moral theology in an emerging technoculture. Waters has served previously as the Director of the Center for Business, Religion and Public Life, Pittsburgh Theological Seminary. He is a graduate of the University of Redlands (BA), School of Theology at Claremont (M.Div., D.Min.), and the University of Oxford (D.Phil.).

Clare Williams is Reader in the Social Science of Biomedicine in the School of Nursing and Midwifery, King's College London. Her key research interests include the clinical, ethical and social implications of innovative health technologies, particularly from the perspective of healthcare practitioners and scientists. She has written numerous publications for social science and practitioner journals, and is author of *Mothers, Young People and Chronic Illness* (Ashgate Press, 2002), and co-author of a forthcoming book with S.P. Wainwright, *The Body, Biomedicine and Society: Reflections on High-Tech Medicine* (Palgrave Innovative Health Technology Book Series).

Preface to this Edition

Should we hope for, should we imagine the future perfection of the human? The future under discussion in this book is the human future as offered by new medical practices and technologies. Is the medicalized manipulation of the human body limited only by our imaginations and tools? Should we hope for a transhuman future in which consciousness is regarded as the 'essence' of the human and its transfer to a medium other than the human body is welcomed? Or is such a hope a dark fantasy which will cause us to misunderstand the nature, potential and possibilities of the human? Should we hope for human 'life' to be developed in laboratory contexts? Or are such practices the de-naturalizing of the human? Should we hope for the extension of the length of human life way, way beyond a span of "three score years and ten"? Where should human beings invest their time, ingenuity and wealth in the future – should such investment be to the benefit of medically-insured individuals? Or should investment in health be directed to tackling underlying social problems associated with poor health such as inadequate housing, lack of creative work and high levels of urban pollution? It is to these hard questions that this collection of essays offers some critical responses.

Moreover, it can readily be appreciated that even if these questions are raised by new medical developments they cannot be answered satisfactorily out of medical expertise alone. These medical developments have already been guided by prior conceptions of the human held in the wider culture. And so it is to conceptions of the body in the wider culture than we must now turn to answer these questions. It is not that medicine is posing questions that society must answer. It is rather that medical practices already embedded in a wider society are bringing to the surface questions that the wider society is now forced to answer. From society must come responses to questions arising from society.

To which resources in society shall we turn? Although the reader will find little agreement among these essays, there is agreement on one matter: religious and theological resources are pertinent. That is hardly surprising: the matter of future perfectibility runs deep in Judaeo-Christian traditions in their reflections on personal immortality. Moreover, the healing miracles performed by Jesus have ensured that a ministry to the sick and dying lies at the heart of Christian notions of hospitality. Add to this the efforts over the last 100 years or so to 'socialise' Christian interpretations of eschatology and we have a convergence of concerns that have obliged theology to offer a critical assessment of recent medical developments.

Since this collection was first published in 2006, moreover, the issues around the future of medical practices and human perfectibility have continued to surface at the boundary of theology and public life. If anything, the debate has intensified.

One of the reasons for this intensification is that the ethical and moral questions around human perfectibility are still unresolved. This is partly because how we think about human identity becomes unsettled when boundary issues are raised, including the boundary between the human and the machine, the human and other animals, or the human and its future. So medical practices come to be viewed in a new light: do new medical practices – and the sentiments that accompany these practices – require new conceptions of the human that are troubling or even inimical to human flourishing? For such practices presuppose a particular understanding of the human and the future of the human. We have here an excellent example of the way in which technological innovation not only alters the human body – and possibilities for the future of that body – but also alters our conceptions of the human body. Do we approve of these new conceptions? How are we to evaluate these new medical practices? And what are the differences and similarities between the techno-futures we are being offered and theology's traditional understandings of future fulfilment?

One of the tasks of this book is to unsettle any complacent truce between medicine and theology for these essays expose the philosophical and political undertow shaping the way human identities and human futures are conceived. Yet while the orientation of many of these essays is eschatological, they also consider how far a theological account of creatureliness still makes sense in such a context: can an understanding of human lives as gifts from God still be persuasive under such pressures for change?

Moreover, is there scope for theology to resist some of the drives behind modern medical practices by uncovering their rootedness in forms of philosophy that may be hostile to religious belief? The possibility of the human filtered through enhancement, posthuman and transhuman discourse continue to attract debate in ways that remain unresolved. This is not simply a matter of defining what these terms mean, but beyond this, the kind of cultural landscape that a transhuman world conjures up, and its influence on cultural acceptance or not of particular medical practices.

As these pressures mount, there is an even greater need for conceptual clarity about human nature, the human condition, and what culturally might be acceptable or not from a theological point of view. We can ask ourselves, for example, if the idea of a trans-historical human nature is defensible or not, given the medical possibilities on offer. Further, in the cultural shadow of the postmodern views of the self and its meaning, doubts about attempts to disengage and disembody that self from particular supposed cultural and gendered restraints come to the surface once more in posthuman and transhuman dreams about the future.

Of course, not all contributors to this volume reach the same conclusion, and the variety expressed theologically shows the possible alternative positions that could be adopted. Some opt for accommodation to new medical and other practices and the task of theology is to welcome and endorse such practices, albeit within certain responsible limits. Others are much more resistant and argue against the trajectory of modern technological medicine, believing that it has succumbed to a particular philosophical view that is alien to Christian belief. Others again take a more intermediary position and argue that theology has a place at the table of public deliberation. Our response to such issues will depend, of course, on presuppositions about technology as such and its relationship to the human.

As editors we are content to let this variety stand and let readers of this book come to their own conclusions about the cogency of the arguments in each case. It is our intention that this book opens up the debate in ways that will inform intelligent discussion of these complex issues. For we stand at a threshold when clear thinking on these topics has become increasingly vital in order to inform policy and practices in different jurisdictions. There are, in other words, important political agendas at stake when medical technology is under discussion. Theologians are often tempted to keep to a particular specialism within their own traditions and by doing so avoid some of these difficult and contested questions. But the editors of this volume hold to the view that theology has a public duty to contribute to such debates about the future of the human. In other words, it is not sufficient just to rehearse or critique the dogmas of the religious past without noting the usefulness or otherwise of their particular insights for understanding present dilemmas. In issuing this as a paperback volume we hope that this theological contribution to a public debate will now be accessible to a wider audience. We are therefore especially grateful to Tom Kraft of T&T Clark for facilitating the publication of this paperback edition.

Celia Deane-Drummond and Peter Manley Scott, September 2009.

Introduction

Future Perfect? Or, What Should We Hope For?

Peter Manley Scott and Celia Deane-Drummond

This collection of essays emerges from a colloquium that met at St Deiniol's library in June 2005. Its intention was to bring together a group of scholars who had a particular interest in medical practices and their future development, be it from a theological, medical, philosophical, sociological or ethical point of view. A major theme was the possibility of shifts in medical practice towards engaging with human perfectibility: how far might this be embedded in cultural practices and should this development be endorsed from a theological and ethical point of view? The following introductory remarks help set the scene of the themes that emerge in this book: Is human identity in some sense being changed by current medical practices? Are these changes morally significant? By which ethical criteria should we evaluate these changes? These are among the questions that are raised by recent developments in medical science. As such, theology is driven to reconsider its understandings of (1) the human, (2) medical practices and (3) eschatology. Let us consider these three in turn, from a theological perspective.

We may note straight away that theology here faces a twofold temptation: either to adopt an attitude that refuses technological change or to seek over-hastily to baptize technological changes. In other words, what is at stake here is what is meant by the *givenness* of creaturely life. Should part of theology's effort at understanding these changes be in the service of preserving the human from such change and warning of the negative consequences of such change? Does the claim that the human is God's creature recommend a strong caution in the face of such change? Or, should God be understood as actively engaged in the capacity of the human to humanize its context, including its own body? That is, may theology endorse change where it seems to be consistent with Christian perspectives, on healing, for example? (Part of the issue here is whether there ever can be a general answer to these questions.)

Moreover, which understandings of God should be privileged in our attempts to understand both the givenness and future of life? Who is this God, how does this God act, and to what purpose? Indeed, to each of these questions, we may present an anthropological correlate: the identity of the human, the nature of change and the *telos* of humanity. We are returned once more to the themes of human identity, practice and eschatology.

1. *Medicine and Human Identity*

As the first section of this volume indicates, we need to engage with a range of
Perspectives on Humans in the context of advances in medical practice. That is,
before we ask about how we should act, we may ask: are there now new moral
phenomena or tendencies that we must take into account? The issue of new perspectives is raised by the emergence of terms such as enhancement, posthuman
and transhuman. These terms are not jargon but function as attempts to answer
the question as to whether there are now new anthropological-moral matters
that require our attention. The distinctions between therapy, enhancement and
transhumanism – however slippery – are attempts to explore different meanings of medical practices and medical technologies. Ted Peters addresses these
differences in his essay. We are familiar with the notion of *therapy* as a medical
practice: 'healing, the addressing of a pathology for the purposes of restoring
health'.[1] *Enhancement* refers to 'medical measures that improve an individual's
functioning or improve the human species beyond what had previously been
thought to be its norm'.[2] *Transhumanism* may be understood as drawing from
a different anthropology in which cognitive function is privileged as the site of
human identity; the task of transhumanism is therefore to find a different medium
than our bodies in which our cognitive functioning may continue (indefinitely?).
Posthumanism, a somewhat broader term, is related to transhumanism in that
some strands in the discussion of the posthuman suggest that the 'improvement'
of the human may be possible in far-reaching ways through the extension or rede-
velopment of humanity through its technology.

Nor should these matters be understood as issues for consideration only in
the twilight world of postmodernity. In a strange way, the unencumbered liberal
self – 'the self at the centre of liberal humanist thought, which is supposed to be
capable of being autonomous, rational, and centred, and somehow free of any
particular cultural, ethnic, or gendered characteristics'[3] – remains present in this
debate: technology will free the human self from unhappy encumbrances such as
the body. Indeed, as some posthumanists put the matter, we may be entering a new
phase of human evolution.

In turn, these technological changes raise the question as to whether or not
human nature is being changed. For in this discussion the term 'identity' may be
understood as a placeholder: it presents – but does not answer – the question as to
whether we are speaking here of changes in human nature. If, as has often been
pointed out, human nature has been informed by meanings of the human body and
its inviolability, the mixing of human body and technological machines unsettles
this inviolability. What, then, of a static human nature? Or, by contrast, should we
say that however human nature is to be accounted for, technological change does
not impact upon it? What is by nature human thereby remains unaffected although
the human condition may be affected. On this interpretation, as Gordon Graham

---

1   Ted Peters, 'Perfect Humans or Trans-Humans?', this volume, p. 17.
2   Peters, 'Perfect Humans', p. 17.
3   Christopher Butler, *Postmodernism* (Oxford: Oxford University Press, 2002), p. 59.

enlarges on in his essay, technological changes impinge upon our condition but not on our nature.

Most of the essays collected here assume that human nature is in some way being changed. (Of course, as Søren Holm points out, this is not the same as claiming that the concept of human nature can or should be dispensed with.) Having said that, there is no agreement on whether theology should welcome or resist such a change or whether theology should follow the line of critical accommodation to technological change. Mostly, the judgement regarding change is implicit rather than explicit. What sort of judgement is involved here in coming to a decision about the 'essence' of this change and its impact on our understandings of human nature?

The vital issue here is the character of the change and its relationship to understandings of human nature. For, as Willis Dulap notes, 'With nature thus transformed and subject to further such transformations, the idea of human nature is cut loose from any trans-historical mooring – as least in so far as this mooring might have any substantive content.'[4] If, for example, biotechnological change means the 'end of nature' in its traditional meaning as that which is other to humanity, what knock-on effect is there on our understandings of human nature? In other words, is the change in non-human nature of such depth that trans-historical understandings of the human – which in turn support the givenness of human nature – are fatally weakened? Such a weakening would mean that the appeal to 'human nature' is rendered much more difficult. And, in turn, as Gareth Jones and Gordon McPhate discuss in the context of neuroscience, the matter of human identity becomes more malleable and perhaps subject to conscious human decision and control.

With regard to the issues presented in this volume, we are faced with some choices regarding the consideration of human identity. These choices – presented by philosophers as well as theologians – are nicely reported in the essays that follow. The chapters by Elaine Graham and Søren Holm argue from rather different perspectives – the theological and the philosophical – as to the changing meaning, and persistence, of concepts of human nature.

Not least, the theologian must enquire whether theological resources are to be placed in the service of shoring up meanings of identity in order to affirm the existence of human nature. If the theologian takes this route, the issue of method is raised. For how is the theological voice to be heard in a public sphere that seems to be increasingly indifferent to Christian wisdom? (Of course, modern states may in order to cover their own moral nakedness appeal to the religions at need; this is not the same, however, as constructing a public version of practical reason and moral deliberation to which Christianity is a contributor.) The essays presented here – especially those by Michael Northcott, Gareth Jones, Neil Messer, Maureen Junker-Kenny, Celia Deane-Drummond and Brent Waters – offer

---

4 Willis Dulap, 'Two Fragments: Theological Transformations of Law, Technological Transformations of Nature', in Carl Mitcham and Jim Grote (eds.), *Theology and Technology: Essays in Christian Analysis and Exegesis* (New York and London: Rowman & Littlefield, 1984), p. 231.

differing responses to this matter. These essays range from a sceptical position that sees Christianity as in profound ways in antithesis to modern medical practices to those that seek a more accommodating position.

2. *Medicine and Technology*

The essays presented in the second section, *Medicalized Humans*, address a cluster of concerns concerning the situation of the human under these new medical practices. What is happening to the human at the intersection of society, technology and medicine? In this section scientists and theologians grapple with this issue.

As human identity raises the matter of human nature so the issue of technological practices in medicine raises the matter of the situation of technological change. That is, in order to answer the question as to how the field of moral action is structured in this new and contested area, we are caught up in the politics of technology, including the politics of the meanings of technology. In other words, technological practices are always present through networks of power. As such, the interpretation of technology is understood in relation to those philosophies of science that understand scientific practice as a contested area. Such analysis contends that science itself is not as neutral as it appears but is bound up with a modern, Western, dominating project of control.

Likewise, technology is present in this already structured field and makes its own contribution to that structuring. Jones' and McPhate's chapters discuss particular medical practices that rely on recent neuro-pharmacological and other developments in medicine, so designed to alter human functioning. Yet, as McPhate points out, there may be ambiguities concerning the benefits of such treatment. Transhumanism offers a more general example of this question: who precisely will benefit from this hugely expensive –ism, and – more to the point – from which tasks will funds have to be diverted to try to make the transhuman dream come true? Can this transhuman dream only come true by means of the deepening of the impoverishment of the poor? Yet this level of interaction – the interaction between society and technology – is not the only level. As a number of the essays presented in this volume suggest (Northcott, Messer, Deane-Drummond, Waters), medical practices themselves provide their own political context and pressures for change. While it may not be recognized by most of those engaged in current medical practice, the drift towards 'posthuman' technologically based enhancement has been enabled by a slow and largely uncritical absorption of transhumanism by medical science. Rather than considering them after the event, it is important to engage with those changes that are on the cultural horizon now, prior to concrete decisions about practice.

Moreover, there are the meanings of technology itself. These meanings are often secreted only implicitly and relate to wider understandings of the modern project. Theology thereby needs to grasp two 'practical' issues as these relate to technological change: (1) the meaning of technology, and (2) the relationship of Christianity to modernity. With regard to the meaning of technology, there is an important distinction to be made between technology understood as somehow beyond human control and technological change as revisable by human beings. Accounts of technological determinism emerge at this point in which a substantialist or essentialist account of technology is understood as emerging in history yet somehow does not pass through human hands. In other words, modern societies can only respond to the technological imperative; these societies must 'ride the tiger'.

Although technological determinism is a complex position we do not need to consider the details here in order to notice a fundamental point: if we accept some version of technological determinism then the field of moral action is structured in a specific way. On account of that structuring, theology may hesitate to approve technological changes as these changes – as determinist – relate uneasily to Christian understandings of God's providence. Put differently, determinism and love do not fit easily together.

Of course, a non-essentialist account of technology may be preferred. In this version technology may be understood as a force or imperative that easily escapes human control. However, such an escape is not constitutive of the meaning of technology. It is not necessarily the case that technology escapes human control. For example, there always remains the possibility – however difficult to render actual – of the democratization of technology. In such a scenario, technology can be directed to socially established goals (such as justice). The field of moral action is not pre-structured by technology such that technology eludes human decision-making and control. Furthermore, some account of invariant human nature may be offered as a way of showing that true technological change always has a human correlate.

It is easy to see how these two approaches are linked to understandings of modernity. More precisely, we are in the realm of considering how Christianity relates to modernity. What is the correct theological option at this point? Some of the contributors to this volume (Northcott, Messer, Waters) are influenced by 'postliberal' tendencies in theology and thereby are sceptical of one trajectory of the modern project that has sought the separation of humanity from nature and the control of the latter by the former through technology. (The writings of Dupré and Latour are pertinent.[5]) Other contributors (Ted Peters in theology, Ulf Görman in philosophy) are not so hesitant and wish to explore in a careful and nuanced way the benefits that may emerge from medical technologies in their contribution to human flourishing and just societies.

We can visit some of the theological commitments operative here. Postliberal strategies include drawing on Christian resources regarding naturalness. In other words, the doctrine of creation is appealed to in order to recommend that certain technological practices are somehow alien to human creatureliness. (Northcott's essay offers a good example of this approach.) The notion of life as a gift and human responsibility towards life are common features of this strategy. Alternatively, the future destiny of the creature in God may be appealed to in order to show that the modern desiring that structures aspects of contemporary

---

5   See Louis Dupré, *Passage to Modernity* (London and New Haven: Yale University Press, 1995); Bruno Latour, *We Have Never Been Modern* (London: Prentice Hall, 1993).

medicine is fundamentally disordered. Again, an effort may be made to show that medicine has sold its soul to the modern project; Christian tradition is deployed to show how deeply mortgaged to the modern project contemporary medicine has become. These approaches may draw on the notion of human beings as made in the image of God. (The essays by Northcott, Messer and Waters are recommended reading here: Northcott and Messer concentrate on medical practices in theological perspective whereas Waters focuses on the different futures projected by posthumanism and theology.)

A different set of strategies may draw on the same set of resources but configure them differently. What is natural to humanity is not set in opposition to technological change. As the essays by Jones and McPhate make clear, the desire to overcome suffering is deeply embedded in Christian tradition and remains an important soteriological value. It behoves Christianity to explore carefully and consider deliberately in what ways modern medicine continues to meet (and expand?) this soteriological value. In rather different ways, Görman and Deane-Drummond offer examples of this approach. Of course, Görman's approach is philosophical yet it is a philosophy in the service of Christian interests whereas Deane-Drummond combines philosophical and theological interests. To make matters more complicated still, a hybrid or combination approach may be adopted: for example, an important modern theme such as individuality, resourced by post-Kantian philosophy, may be developed in order to resist certain medical developments (Junker-Kenny's essay provides an example of this approach). Nonetheless, all three are interested in the diversity of eschatologies present in the debates on human ageing and perfectibility.

How is the field of moral action structured for theology? In the context of contemporary medicine, this proves to be a difficult question! In these essays there is no agreement on how this field is structured. Technological practices are always practices of power in which particular meanings of technology are vested. The meanings of technology are always established in theology by reference to Christian resources and the relationship – however differently understood – between Christianity and the modern project (in its variety). Central here is the meaning and valuation of technological change.

The two themes that we have been discussing so far are intertwined. What is the human? What is technology and what is technological change? Answers to these questions cannot be established independently. Moreover, there remains the matter of eschatology: the issue of the *telos* of the human, mentioned above. The next section treats this matter directly.

3. *Future Perfect*

Medicine has traditionally concerned itself with the healing of the sick, of attempting to follow diagnosis with prognosis and therapy, of working towards a future that seeks to alleviate the multiple sufferings of humanity, at any stage from conception until death. Preventative medicine has concerned itself with prevention of the incidence of disease, of avoiding those situations that lead to suffering in the first place. Underlying both practices is a model of the human that presupposes suffering is an evil to be avoided, and accordingly medical practice works consistently to avoid such suffering.

Competing goals in medicine, such as the preservation of life, have in the past obscured this overriding goal to avoid suffering. Nonetheless, the hospice movement has reinforced medicine's original aim against suffering, so that in practice those who are dying are offered palliative drugs that secure the minimum level of pain, even if it means a relative shortening of life. More recently still, there are those who argue that a life that is profoundly and unavoidably full of pain and suffering is one that is not worth living. On this basis there has been added pressure either to eliminate that life as intolerable existence, with an increasingly vocal lobby supporting assisted dying and euthanasia, or avoid the creation of life full of suffering at the start through screening technologies that remove those embryos most likely to develop serious diseases.

Genetics-based medicine has added another dimension to this scenario by seeking to discover the underlying genetic causes of disease, with the eventual aim of modifying those genes. The hope here is that the person concerned becomes completely free from the suffering of that disease, including the possibility of passing that disease to the next generation. Of course, those who aim to treat human subjects through the application of genetic medicine provoke less controversy compared with those who argue for inherited genetic change through embryonic or gamete manipulation, for the latter impinges on not just the resulting person, but on future generations as well. Junker-Kenny's chapter raises questions about the philosophical justification of the drive towards perfection, arguing that there are important philosophical questions about such trends in the case of genetic enhancements. Following Habermas, she argues that the latter allow for an unacceptable domination of the child by its parents who must presume to know its best interest. It is also misguided in the sense that truly human potential comes through knowing and understanding human limitations so that the drive to genetic perfection is inappropriate as a life project.

The public discourse about embryonic stem-cell research is similarly one that is loaded with the language of utopian hope alongside more apocalyptic fears about the future. Jenny Kitzinger and Clare Williams' chapter illustrates well how far the public controversies about stem-cell research play on different perceptions of the future, combined with a rhetoric of compassion either for those likely to be recipients of the technology, or towards embryos 'used' in order to generate stem cells. Genetic medicine has also added another dimension to the discussion in that those who have only a *propensity* for a disease are also considered to be 'patients'. Even though such patients are free of specific symptoms now, they may suffer in the future.

Ageing is another area worth considering in this context: those who are elderly – with its associated discontents – in some cases become targets for medical practice, with the eventual goal of alleviating not just the diseases that are concomitant with old age, but also delaying the ageing process as such. Görman's chapter draws out the different strategies in biomedicine that try to come to terms with ageing and death. These strategies are to 'normalize' ageing, that is prevention of premature death, to 'optimize' ageing, or maintain good health until death, to 'retard' ageing and postpone death and, finally, to 'eliminate' or overcome ageing and death altogether. For him, the greatest risk of tendencies towards life extension through enhancement lies in its potential for dehumanization. Similarly, Waters' chapter is particularly critical of medical practices that seem to have forgotten the original drive towards caring for the sick in favour of eliminating that sickness in the first place. He sees such a shift as an unwarranted pandering to trends implicit in what has come to be known as 'posthuman' philosophy.

Cultural goals towards perfection, especially of human bodies and cultural models of beauty, also become superimposed on medical practice, so that technologies are used not just for the elimination of physical suffering but also to pander to the desires and wishes of those who desire perfect bodies, with a cluster of narratives around the elusive elixir of youth. Such interventions may be justified on the basis that the psychological suffering of the individual is too great to bear and has a damaging effect on other physiological functions. There may, of course, be instances where cosmetic surgery is entirely justified on this basis. However, the free-market economy in medicine means that there are few constraints on medical practice in this regard, other than the willingness to pay for treatment with due consent by the patient as to the inherent risks associated with such practices.

Overlying all such practices are cultural assumptions about the need for perfection in bodily function, both free from disease, and expressive of cultural norms of beauty. Redemption in such a secular context is redemption from suffering to such an extent that *any* form of suffering becomes a target for medical practice. In addition, enhancement technologies seek to go beyond simple treatment of disease and aim towards an ideal goal set by the norms of society. While there has been much discussion about how far it is possible to distinguish enhancement and therapy, there is a creeping tendency for more and more enhancements to be justified on the basis that they are 'therapeutic': meeting the purported psychological needs of 'patients'. This is a secular soteriology, a form of salvation achievable through medicine, but one that has not only overreached its original goals of eliminating suffering understood in terms of disease, but now also has aimed for forms of perfection as if these are necessarily normative for society. The slip from therapy to enhancement in the context of ageing is an area that Deane-Drummond considers in her chapter.

What might be the response of Christian eschatology to such trends? In the past, models of Christian eschatology have assumed that physical perfection in this life is unattainable, and all attention is focused on future perfection of the inner person. Moreover, while there is a Christian injunction towards healing of the sick, and such healings are in one sense 'signs of the kingdom', an overriding temptation for eschatology is to focus on the inner spiritual life with God as that which is redeemed in the future. However, more contemporary discussion of eschatology looks to the way in which the life of the future envisaged as eternal life both impinges now on present reality, but is also profoundly 'not yet' visible. The question, then, arises regarding realized eschatology: how might we discern the way the Holy Spirit is presently moving in a particular community?

First of all, such eschatology has little to do with the drive towards physical perfection implicit in contemporary medicine. Secondly, it avoids a dualistic splintering of physical from spiritual in which the body is devalued in some way, or merely treated as an empty vessel that is a container for spiritual life. Thirdly, in the context of Christian belief it is grounded in an appreciation of the signifi-cance of the Sonship of Christ, including incarnation into material reality, his life, suffering, death and resurrection.

Taking such eschatology further, we might ask ourselves in what sense in the movement of eschatology does the world dwell in God, or does God dwell in the world?[6] It is possible to argue for both approaches as having validity. The first approach is one favoured by theologians such as Hans urs von Balthasar, and the Eastern Orthodox tradition, and puts most emphasis on the sharing of created beings in the life of the Trinity, through participation in the divine life of perichoresis.[7] Redemption consists of creatures sharing in that divine life, a dwelling of the finite in the infinite. In Deane-Drummond's chapter this dimen-sion of Christian eschatology comes to the fore, since it shows the way in which perfection is an integral component of Christian hope. Such perfection is not so much related to bodily function but is expressed through the grace of God as expressed in the virtues. Yet, a second approach also has its validity, namely that God enters into the world through divine condescension, an immanent presence in the world such that God shares in the suffering of creation. A Christological approach allows for both movements, for while the first puts more emphasis on the gift of the Son in terms of the Trinitarian life, the second puts more emphasis on the incarnate human Christ as one who empties himself of divine power, sharing in the process of suffering through a process of kenosis. This second movement is also in tune with Jewish expectation about the 'new Jerusalem'; it is a '*finitum capax infiniti* – a finitude that embraces infinity'.[8] While Moltmann is now more inclined to favour the second movement of God indwelling the world, finding 'gnostic' tendencies in the former indwelling of the world in God, there is ulti-mately no contradiction between the person who abides in love in God, and God who dwells in him (1 John 4.16). Deane-Drummond also points to the importance of considering the whole life of creation as integral to Christian eschatology, so that human life is not detached from the social, political and ecological aspects of existence. The question now arises: how does this eschatology envisaged here serve to critique those false desires associated with the perfection of the human from a secular perspective? In other words, what should we hope for?

4. *What Should We Hope For?*

As we look into more practical consequences of experiences of God and how these intersect with eschatology as envisaged above, this serves to shed light

6   See, for example, J. Moltmann, 'The World in God, or God in the World?', in R. Bauckham (ed.), *God Will be All in All: An Eschatology of Jürgen Moltmann* (Edinburgh: T & T Clark, 1999), pp. 35–41.

7   See H. Urs von Balthasar, *Theo-Drama: Theological Dramatic Theory. V. The Final Act*, trans. Graham Harrison (San Francisco: Ignatius Press, 1998).

8   Moltmann, 'The World in God', p. 40.

on and contrast with the image of the future presented in medical soteriology. If Christian hope is also that hope coming to meet humanity and creation, an *adventus*, rather than that emerging out of the present, a *futurans*, then once more the contrast with existing hopes becomes evident.[9] Experiences of life that entail suffering give opportunities through Christian hope for new beginnings, the courage to start over again.[10] Junker-Kenny's chapter similarly reminds us that human perfection in a Christian eschatology has less to do with biology and more to do with seeking to express the image of God in a way that also expresses diversity, rather than uniformity. Moreover, such a transition comes through a Christological reference, so, drawing on Schleiermacher, it is the *God-consciousness* found in Christ that provides the model of human perfection understood from a Christian perspective.

This is not a glorification of physical suffering, but a recognition that attempts to avoid all suffering are misguided, for they fail to tackle the reality of human life as limited, fragmentary and often full of guilt and psychological dis-ease. The message of Christian eschatology is still one of hope in spite of misguided attempts to save ourselves through our own effort and will. Moltmann describes this as a time of patience in God, preserving the world and not destroying it in spite of its self-contradictions. He suggests that God 'takes on himself the dissonance between the world's creation and its corruption, so that in spite of its corruption the world may live'.[11] Christ's death and resurrection expresses this hope concretely, for it is a sign that God's future has already begun. Waters' chapter argues for a deeper appreciation of Christology in understanding human nature, so that the move towards any so-called 'posthuman' future in modern medicine is strongly resisted.

A further aspect is a theme that has recurred throughout this volume, namely that the attempt to aim at immortality through forms of posthuman medical practices is misguided. This might be seen, also, as a secularized version of theological attempts to separate the soul from the body, for it amounts to a dehumanizing of what it is to be human in the concrete sense, eventually aiming at artificial minds that are simply silicon chips, a disembodied brain. Just as Moltmann is critical of that dualistic Platonic philosophical tradition that speaks of a disembodied soul, similarly, the core question for humanity in reflection on the future should not so much be the distinction between soul and body, as the conflict between love and death.[12] He argues that God's love in relationship for humanity is sustained through death, for if it were not so, death would be greater than God. As God experiences human life, it becomes eternally immortal, even though death is experienced by humanity as the limit of human life. On the one hand humanity cannot hold on to life, for everything passes away. This is the mistake of medicalized technologies where they seek to aim for life extension and beyond to posthuman futures. On

---

9   For discussion of this see J. Moltmann, *The Coming of God* (London: SCM Press, 1996), pp. 25–26.

10   J. Moltmann, *In the End – the Beginning* (London: SCM Press, 2004), p. x.

11   Moltmann, *In the End*, p. 41.

12   This recurs in his writing, most recently *In the End*, p. 105.

the other hand, everything remains in God in as much as human life remains in God. Eternal life, for the Christian, is not so much an extension of human life, but what Moltmann calls 'an eternal livingness', understood in terms of intensity of experience.[13]

Such intensity is one expressed most fully in love. In addition, where humanity seeks to find such fullness of life in themselves alone, they will inevitably be disappointed. Christian hope of ultimate fulfilment, of being present in God, relativizes the striving that presumes to replace Christian hope with a secular version of immortality. At the same time, the indwelling of God in creation speaks of a care about the present life which encounters human suffering, and comes alongside those who express these virtues of love and compassion in the face of suffering and death. In addition, medical practice needs to be alert not just to the trends in society towards perfection and enhancement, pandering to those wants, rather than needs, but also recover its goal of being dedicated to the service of humanity and the common good. Such practices may take the shape of a reformulated understanding of the virtues, embedded in medical practice, as Messer has suggested, or those virtues, such as prudence and justice, that are particularly relevant where decisions about policy are complex and there are competing goods, as Deane-Drummond argues. Christian virtue, understood in an eschatological perspective, offers hope for the future, but at the same time refuses to avoid suffering and facing the harsh reality of death. Wisdom, for example, as virtue is a positive goal in that it speaks of the vitality of the relationship with God, but at the same time it is also counter-cultural through the wisdom of the cross.

More importantly, perhaps, the absorption of transhuman or posthuman goals is neither necessary nor desirable in order for medicine to be relevant to the needs of contemporary society. As Junker-Kenny has indicated, there are philosophical as well as religious reasons for resisting the drive towards perfection that is slowly creeping into modern medical practice. Such philosophical reasons include the issue of human identity, but this is then deepened further by reflection on religious insights in Christian theology.

All the authors of this volume are supportive of the overall aim of medicine to heal the sick. There are differences of opinion as to how far other agendas have crept into medical practice, and how far such agendas represent the rhetorical expression of media-led debates that are detached from grassroots experience. All authors are attuned to the social context in which medical practice is taking shape, though there are differences in opinion as to how far medicine is informing or being informed by this context, and to what extent medicine is now departing from its traditional goals. In particular, in what sense is medicine seeking to alleviate what was once thought to be given aspects of the human condition?

The mixture of backgrounds of authors for this volume includes those with primary training in medicine, philosophy, sociology, ethics and theology. It is our hope that this mix will be stimulating for the reader and encourage him or her to take these discussions further in the light of their own particular interests and concerns. This book offers, then, a starting point in discussion about the particular

13   Moltmann, *In the End*, p. 152.

hope engendered by medicine in the light of dreams of perfection and its critical appraisal on a number of different levels – including, most importantly, the levels of ethics and theology and their image of the human.

Part I

Perspectives on humans

Chapter 1

PERFECT HUMANS OR TRANS-HUMANS?

Ted Peters

Perfection through science: is this a realistic goal? Whether realistic or not, should the striving for perfection provide an ethical vision or guide? For John Wesley, Christian perfection is possible in this life through the work of the Holy Spirit within a loving heart. Can medical science replace the Holy Spirit?

What counts as perfection? Weslyan Methodists and Christians more generally aim at cultivating a loving and caring disposition: 'pure love reigning alone in the heart and life, this is the whole of scriptural perfection'. [1] What would count as perfection for medical science? Would the perfection toward which medical therapy aims be a state of good health? Would the perfection toward which genetic or other physical enhancement aims be excellence, or superiority? Would transhumanism give up on human perfection as too restrictive and strive for a supra-human perfection?

What role does a vision of perfection play? In Christian perfection, the vision of a loving heart draws the person of faith toward increased caring, toward transformation from selfishness to selflessness. In medical therapy, the vision of good health directs the steps to be taken by doctor and patient. In considering enhancement, the vision or model of the excellent or superior person suggests genetic or other biological alterations that could produce this effect. In assessing the emerging concept of transhumanism, visions of good health and enhanced capacities carry our imaginations beyond what we previously thought were within the range of finite human existence.

With these questions and prospects in mind, I would like to examine carefully three forms of visioning that lead to transformation through medical science: therapy, enhancement and transhumanism. All depend upon an image of who we are and who we would like to become. Enhancement and transhumanism each add an element of the fabulous, an element of excellence or superiority that takes us beyond good health. Do they require an implicit image of perfection? Or, is the mere striving for partial betterment sufficient to energize and direct the relevant medical sciences?

The track we will follow will be that of genetic or other biological interventions as they pass through therapy, enhancement and transhuman projections. When we switch to the ethical and theological tracks, we find little or no controversy surrounding therapy; mild caution regarding enhancement; and rejection of at least the extreme forms of transhumanism.

---

1   John Wesley, *A Plain Account of Christian Perfection* (London: Epworth, 1952), p.52.

One of the problems associated with enhancement and transhumanism is that they risk promising too much. They promise to make us into fabulous human beings, in the latter case even something posthuman. They promise transformation, and transhumanism even promises transformation leading to a new level of perfection. Tacitly, they risk assigning science the job of providing the equivalent of salvation. Our risk is that our generation might be tempted to ask of science what only God can deliver. God may be able to deliver perfection. Science cannot.

The God of the Bible is a God of transformation. The Easter resurrection of Jesus announces God's promise of a radical transformation yet to come: the eschatological new creation. In anticipation of the new creation, Jesus performed miracles of healing. We might call his healing 'eschatological therapy'. Those healed experienced transformation as today's effect of tomorrow's renewal. Built right into the Christian world-view is the vision of a coming new creation accompanied by a ministry of healing in today's broken world. Any Christian ethic should embrace transformation aimed at healing, to be sure.

As we move from eschatological vision to bioethics, we need to distinguish between promises peculiar to God and those we can realistically assign to medical science. Ultimate transformation and perfection belong to God's agenda. Striving for wellbeing and even flourishing belong on the human agenda, and this includes advancing scientific research as a means to that end. Knowing the difference between a vision of perfection and our appropriate striving will require wisdom and prudence.

Wisdom and prudence are what Celia Deane-Drummond offers. Her virtue approach to ethics is 'eschatological in orientation; it explores through a particular theological telos, what might be the good end for humanity'.[2] With the ultimate end envisioned, she then commits herself to honouring wisdom, which is 'an expression of the eternal mind of God, while at the same time affirming what can be known in creaturely existence'.[3] From wisdom emerges her virtue ethic. She follows Thomas Aquinas by placing prudence in special relation to the other cardinal virtues: justice, fortitude and temperance. Prudence is the application of wisdom to human affairs; and ethics counts as such a human affair. 'The ability of prudence to be still, to deliberate well, is a quality desperately needed in the frenzied search for new methods and techniques in biological science that are considered to have particular usefulness for humanity.'[4]

My proposal is that Christian deliberation over speculations regarding the next advances in neurocognitive research affirm at the level of assumption that transformation on behalf of improved human flourishing is attuned to God's eschatological promise; and, in addition, I propose that we invoke wisdom and prudence in assessing what is realistic about the promises of medical advance. I will conclude that therapy is an uncontroversial divine mandate, that transhumanism is unrealistic, and that enhancement deserves continued ethical monitoring and sorting out.

2   Celia E. Deane-Drummond (ed.), *Brave New World? Theology, Ethics, and the Human Genome* (London and New York: T & T Clark International, 2003), p. 235.

3   Celia E. Deane-Drummond, *The Ethics of Nature* (Oxford: Blackwell, 2004), p. 20.

4   Drummond, *Ethics of Nature*, p. 14.

1. *From Therapy to Enhancement*[5]

For the sake of this discussion, I would like to distinguish three terms, *therapy*, *enhancement* and *transhumanism*. The first two are commonplace in medicine and bioethics. By 'therapy' we mean healing, the addressing of a pathology for purposes of restoring health. By 'enhancement' we refer to medical measures that improve an individual's functioning or improve the human species beyond what had previously been thought to be its norm. Therapy is a response to a pathology, whereas enhancement initiates an improvement without reference to a pathology.

Therapy restores health. What is meant by health? The World Health Organization holds that 'health is a state of complete physical, mental, and social well being and not merely the absence of disease or infirmity'.[6] The late Pope John Paul II offers a similar definition. 'From a Christian perspective, then, health envisions optimal functioning of the human person to meet physiological, psychological, social, and spiritual needs in an integrated manner.'[7] Although such definitions may open themselves to ambiguity, the end or goal of therapy or cure gains some helpful specificity.

Even though good health may be pursued with passionate hopes for a cure, and even though therapy's achievement results in joy, good health is not a form of enhancement. To be healthy, we assume, is to exist at the norm, to live as our bodies were originally designed to live. Despite the joy good health brings to the healthy person, we would want everyone in our neighbourhood to benefit from the same level of wellbeing.

Enhancement is distinguished from therapy here because it involves efforts to make someone not just healthy, but better than healthy. More than offering a cure, enhancement optimizes attributes or capabilities beyond what good health requires. The goal might even be to raise an individual from standard up to peak levels of performance. If therapy via genetic intervention would bring an individual up to what is average or normal, then enhancement would bring an individual beyond the average or norm up to a level of excellence above others. Eric Juengst defines enhancement this way: 'The term *enhancement* is usually used in bioethics to characterize interventions designed to improve human form or functioning beyond what is necessary to sustain or restore good health.'[8]

The US President's Council on Bioethics (COB) defines *human gene therapy* as directed genetic change of human somatic cells to treat a pathological situation, a genetic disease or defect.[9] By *human genetic enhancement* the COB refers to

5 My background study of genetic therapy and enhancement has been pursued as part of a larger research project, 'Altering Nature', at Rice University on a grant from the Ford Foundation, with Andrew Lustig as Principal Investigator. The team researching enhancement issues includes Karen Lebacqz, Crawford Cromwell, Bernard Lo and Estuardo Aguillar, along with me serving as chair.

6 http://policy.who.int/cgibin/om_isapi.dll?hitsperheading=on&infobase=basicdoc&jump=Rel ations%20with%20NGOs&softpage=Document42#JUMPDEST_Relations%20with%20NGOs.

7 Pope John Paul II, 'The Ethics of Genetic Manipulation', *Origins* 13.23 (1983), p. 385.

8 E.T. Juengst, 'What Does *Enhancement* Mean?', in Erik Parens (ed.) *Enhancing Human Traits: Ethical and Social Implications* (Washington, DC: Georgetown University Press, 1998), p. 29.

9 'Staff Background Paper: Human Genetic Enhancement', US President's Council on

the use of genetic knowledge and technology to bring about improvements in the capacities of living persons or future generations.

2. *From Enhancement to Transhumanism*

Beyond enhancement, the open arms of transhumanism seem to be welcoming us. 'Transhumanism is the view that humans should (or should be permitted to) use technology to remake human nature,' write Heidi Campbell and Mark Walker.[10] Through genetic technology, information technology and nanotechnology transhumanists believe the possibility exists for us to greatly enhance the healthy lifespan of persons, increase intelligence, and make ourselves happier and more virtuous. The key is to recontextualize humanity in terms of technology. This leads to a vision of a posthuman future characterized by a merging of humanity with technology as the next stage of our human evolution. *Posthuman* refers to who we might become if transhuman efforts achieve their goals. We are on the brink of becoming more than human, say the transhumanists.

For transhumanists, death should become voluntary. Immortality should become an option. Once we advance to the posthuman future, we will no longer be required to die. Immortality will belong to our definition as posthumans. Now, how do we get from here to there?

One step toward taking us from our present mortal human state to a posthuman immortal state is to build upon a key assumption, namely, that who we are as persons is centred in our brain activity. Transhumanism assumes that intelligence as a pattern of brain activity is the home of our personhood. It is the cognitive code or information pattern that is definitive. We are what we think, so to speak. Therefore, to improve the human situation we may improve on biological function insofar as it enhances cognitive function; and, if necessary, we might even transfer cognitive function into a machine. The goal of transhumanism is to download the contents of human consciousness onto a vast computer network and, via the network, achieve a kind of disembodied yet intelligent immortality. As a machine, we could enjoy perfections that the limits of biology prevent us from attaining.

Anne Foerst explains. She founded and directed the 'God and Computers Project' at MIT's Artificial Intelligence Laboratory in the mid 1990s. She worked in the field of robotics, with AI entities known as 'Cog' and 'Kismet'. Trained in both theology and computer science, she found her MIT colleagues believing 'that humans are nothing but meat machines that carry a computer in their head. As soon as we have decoded the program that runs on the wetware of the brain, we

---

Bioethics (December 2002), www.bioethics.gov/. The term 'enhancement' as currently used in genetics was introduced by W. French Anderson, 'Human Gene Therapy: Scientific and Ethical Considerations', *The Journal of Medicine and Philosophy* 10 (1985), pp. 275–91.

10   Heidi Campbell and Mark Walker, 'Religion and Transhumanism: Introducing a Conversation', *Journal of Evoluton and Technology* 14.2 (August 2005), p. 1. See Nick Bostrom, home page 2005, www. nickbostrom.com/tra/values.html.

can download it into the hardware of a computer and live forever. This is the major belief of a movement that calls itself *transhumanism* or *extropianism.*[11]

The postulated sequence goes something like this. First, AI researchers will simulate human intelligence in a computer, in a robot. Second, humans and machines will merge step by step, replacing portions of our brains with mechanical parts. Third, AI researchers will reduce existing human intelligence to a pattern of information processing and download this into a computer or a robot. This will constitute an evolutionary advance, actually a leap forward that could lead to cybernetic immortality – that is, immortal intelligent life in a machine that gets constant backups.

Ray Kurzweil predicts that by the end of the 21st century human beings will attain cybernetic immortality. Up until now, he says, our mortality has been tied to the longevity of our hardware. So, when our hardware crashes, our thought processes crash with it. When we instantiate ourselves in our computational technology, our software and hence our immortality will no longer be dependent on the survival of our body. Our immortality will be contingent on our being careful to make frequent backups.[12]

How likely is this to come to pass? Noreen Herzfeld, a computer scientist and theologian at St John's University in Collegeville, Minnesota, has tracked earlier projections of AI achievements since the 1950s. The record is dismally poor. Goals have not been reached even now in the early 21st century, despite enormous progress in computer development. No computer yet in existence can be deemed intelligent. 'While computing in general has advanced dramatically in the last fifty years, advances in AI have been limited. Neural networks remain at a level far below the complexity of the human brain. Current research in neuroscience suggests that the workings of the brain are far more complicated than was initially supposed and may not be capturable in neural net technology as we currently conceive it.'[13]

In all three – therapy, enhancement and transhumanism – we employ scientific research and medical technology to intervene in our natural processes to attain a certain end. The end for the first two is improved biological functioning. The end of the third is improved cognitive functioning with or without a biology.

---

11   Anne Foerst, *God in the Machine: What Robots Teach Us About Humanity and God* (New York: Dutton, 2004), p. 43.

12   Ray Kurzweil, *The Age of Spiritual Machines* (New York: Viking, 1999), chapter 6. Ian Barbour is slow to grant the assumption that we can transfer human consciousness to silicon. 'I suspect that it will turn out that conscious awareness requires forms of organized complexity or properties of neural cells and networks that have no parallels in silicon-based systems. I do not think we can exclude the possibility of conscious computers on metaphysical grounds, but there may be empirical grounds for the impossibility of computer consciousness... I am willing to leave this question open.' Ian G. Barbour, 'Neuroscience, Artificial Intelligence, and Human Nature', in Robert John Russell, Nancey Murphy, Theo C. Meyering and Michael A. Arbib, *Neuroscience and the Person: Scientific Perspectives on Divine Action* (Vatican City State and Berkeley, CA: Vatican Observatory and Center for Theology and Natural Sciences, 1999), p.266.

13   Noreen L. Herzfeld, *In Our Image: Artificial Intelligence and the Human Spirit* (Minneapolis: Fortress Press, 2002), pp. 72–73.

3. *Health and Enhancement: A Blurry Line?*

When we compare the first two, we can see that the line between good health and enhancement might be difficult to draw, especially if both seek to optimize functioning in an integrated manner. The distinction is blurry, not sharp. According to the World Council of Churches, 'there is no absolute distinction between eliminating "defects" and "improving" heredity. Correction of mental deficiency can move imperceptibly into enhancement of intelligence, and remedies of severe physical disabilities into enhancement of prowess.'[14]

Despite the blurry line, 'therapy' seems to garner approval while 'enhancement' is greeted with moral suspicion. Francis Fukuyama speaks for many: 'One obvious way to draw red lines is to distinguish between therapy and enhancement, directing research toward the former while putting restrictions on the latter.'[15] Geneticist W. French Anderson holds that 'we should not step over the line that delineates treatment from enhancement'.[16] Both therapy and enhancement require altering our bodies; yet the motive of healing seems distinguishable from that of enhancement.

The line between enhancement and transhumanism can also be blurry. Neuroscientists are currently employing chip lithography to create silicon neuronal chips. These silicon chips have furrows cut in them so that, when implanted in the human brain, actual neurons will grow in such a way that their dendrites and axons interact to create new input and output patterns. The merging of machine and brain could enhance thought, perhaps carrying a person beyond mood alteration or executive improvement to an advance in calculation ability.

Curiously, experiments with computers are going the opposite direction. Because DNA is the single best information storage system yet discovered, scientists are experimenting to see if they can create DNA computers with storage capacities that will leap ahead of anything we have created synthetically. We are technologizing the organic world while organicizing the technological world. Whether inside or outside the human brain, calculation capacity is on the verge of a quantum leap forward.

Bioethicist Paul Wolpe finds these developments astonishing yet promising. 'The point of all this is that we really are becoming some kind of cyborg, some kind of posthuman in the sense that for the first time in history we really are going to incorporate our synthetic technologies into the very physiology of our being – with major, though not necessarily entirely undesirable, consequences.'[17] Arthur Caplan denies that these developments will make us posthuman; rather,

---

14   World Council of Churches, *Manipulating Life: Ethical Issues in Genetic Engineering* (Geneva: World Council of Churches, 1982), p. 7.

15   F. Fukuyama, *Our Posthuman Future* (New York: Farrar, Straus and Giroux, 2002), p. 208.

16   W. F. Anderson, 'Genetics and Human Malleability', *Hastings Center Report*, 20 (February 1990), p. 24.

17   Paul Root Wolpe, 'Neurotechnology, Cyborgs, and the Sense of Self', in *Neuroethics: Mapping the Field*, Dana Foundation Conference Proceedings, San Francisco, 13–14 May 2002 (New York: Dana Press, 2002), p. 164.

he celebrates their enhancement of our humanity. We should applaud 'the new knowledge the brain sciences are providing to try to improve, enhance, and otherwise move toward optimization of our brains'. [18]

Ethicists who anticipate developments in neurocognitve enhancement identify four ethical issues.[19] First, safety. What might be the long-term biological effects of interventions now via either neurotechnology or psychopharmacology? Will today's Ritalin children tomorrow confront premature cognitive decline? Second, coercion. If neurocognitive enhancement becomes widespread, might we confront situations where people are pressured to enhance their cognitive abilities? Will employers pressure employees, or courts pressure convicts? Third, distributive justice. More than likely, the upper economic classes will benefit more from brain enhancement than those unable to afford such medical opportunities. Is this just? Fourth, personhood and intangible values. Will neurocognitive alterations intersect with what we believe makes us persons? Will it make difficult our ability to envision what it means to be healthy and whole? Will it dull us from appreciating and valuing human life even with its imperfections?

4. *Theology and the Ethics of Enhancement*

The challenges of enhancement and transhumanism are knocking on religion's door. Will that door open or remain shut? Heidi Campbell and Mark Walker present transhumanism in terms of its salvific potential. Technology is becoming a rival to religious promise. The transhumanist vision is compelling, they say, because it touches 'on a desire for a life that overcomes the brokenness of this world, a place where pain and suffering are eliminated. This is a longing that is articulated in many religious traditions, those that subscribe to a distinctive eschatological belief in a future where humanity is perfected and transformed.'[20] We are on the brink of a technological eschatology, they assume. What might theologians have to say about this?

Relatively little has been said by Christian theologians and ethicists on the specific topics of brain enhancement through genetic intervention, neurosurgery, or pharmacology, let alone transhumanism. However, the more general question of enhancement has been raised on occasion. We will look here at the question of therapy versus enhancement in Eastern Orthodox ethics, Roman Catholic moral theology, and Protestant thought.

Eastern Orthodox theologians place their ethics within the framework of eschatological perfection, perfection understood as union with God. 'The Christian must place health care, the amelioration of suffering, and the postponement of death within the pursuit of holiness,' writes Tristram Engelhardt.[21] Within such

18   Arthur Caplan, 'No Brainer: Can We Cope with the Ethical Ramifications of New Knowledge of the Human Brain?', in *Neuroethics: Mapping the Field*, p. 97.

19   Martha J. Farah *et al.*, 'Neurocognitivie Enhancement: What Can We Do and What Should We Do?' *Nature Reviews* 5 (May 2004), pp. 421–25, www.nature.com/reviews/neuro.

20   Campbell and Walker, 'Religion and Transhumanism', p. 2.

21   H. Tristram Engelhardt, Jr., *The Foundations of Christian Bioethics* (Liss: Swets & Zeitlinger, 2000), p. 354.

a framework, John Breck can celebrate the potential benefit from genetic engineering (GE) for purposes of therapy; but he is more than merely cautious at the prospect of using genetic intervention for purposes of enhancement and eventually eugenics. 'The line between therapeutic and eugenic techniques is difficult to draw,' he writes. 'While theoretically it is possible for GE to replace *homo sapiens* with a superior *homo novus*, such "enhancement technology" is far in the future... Nevertheless, the potential for "improving" on God's blueprint for human life is such that serious ethical questions must be addressed here and now, by the Churches as well as by public and private regulatory bodies.'[22] If the Orthodox are cautious about biological enhancement, they are likely to be even more reticent when confronted with transhumanism.

Roman Catholic theologians are also less than likely to advance an ethic supporting enhancement or posthumanism. A recent International Theological Commission working for the Vatican's Congregation for the Doctrine of the Faith has published a thorough discussion of Christian anthropology, *Communion and Stewardship: Human Persons Created in the Image of God*. It celebrates what it calls the 'biogenetic characteristics' that allegedly make each person unique. To modify a person's genetic code, regardless of motive, would infringe on a person's identity, says the commission.

> Changing the genetic identity of man as a human person through the production of an infrahuman being is radically immoral. The use of genetic modification to yield a superhuman or being with essentially new spiritual faculties is unthinkable, given that the spiritual life principle of man – forming the matter into the body of the human person – is not a product of human hands and is not subject to genetic engineering.[23]

The logic seems to be this: genetically designing a posthuman is immoral because it is impossible and, if it were possible, it would still be immoral because it compromises human identity.

Significant here is that Catholic caution regarding enhancement would arise not from the technologies themselves, but from an understanding of what human excellence means. Catholics are leery of using medicine for enhancement because efforts at enhancement are so obviously bound up with value choices. A Protestant writing in a Catholic journal, Donal O'Mathuna, says 'In making a person taller medicine promotes the belief that short people are of lesser value and that height is significant in achieving the good life.'[24] Some of the values which might guide enhancement are questionable from a Christian perspective: 'Physical health is not the ultimate priority in Christian eyes.'[25] In short, if the *telos* or proper end of

22   J. Breck, *The Sacred Gift of Life: Orthodox Christianity and Bioethics* (Crestwood, NY: St Vladimir's Seminary Press, 1998), p. 195.

23   *Communion and Stewardship: Human Persons Created in the Image of God*, para. 91. The Vatican, International Theological Commission, Congregation for the Doctrine of the Faith, www.vatican.va/roman_curia/congregations/cfaith/cti_documents/rc_con_cfaith_doc.

24   Donal P. O'Mathuna, 'Genetic Technology, Enhancement, and Christian Values', *The National Catholic Bioethics Quarterly* 2.2 (Summer 2002), p. 283.

25   Mathuna, 'Genetic Technology', p. 284.

human life is the development of a relationship with God and the cultivation of those excellences or virtues necessary for living in community, then even some forms of suffering can be seen as serving those ends. Not all suffering should automatically be removed from human life. 'Attempting to remove the trials and difficulties of life by genetic enhancement might derail the very ways in which God wants to shape our characters.'[26]

Is there wiggle room here? Pope John Paul II did not explicitly forbid genetic enhancement. What needs to be protected is human identity and dignity; and if enhancement could render this protection perhaps it would not be *prima facie* illicit. James Walter comments: 'experiments that are not strictly directed toward therapy but are aimed at improving the human biological condition (enhancement) can be justified, at least in part, on the grounds that the experiments respect the human person by safeguarding the identity of the person as one in body and soul (*corpore et anima unus*)'.[27]

Just how much should we expect from enhancement or related improvements? Might we become unrealistic? Celia Deane-Drummond is sceptical about the use of 'salvific language' on the part of scientists pressing for genetic enhancement of human intelligence.[28] Too much is being promised here. Not only are expectations raised above what reality can possibly deliver; extravagant promises rely on values that are impatient with our humanity, that take us beyond our natural human state. What Deane-Drummond would propose is a virtue ethic drawn from an appropriation of wisdom. 'The theological motif we need to recover as an appropriate response to the new biology is that of wisdom. Such wisdom...offers guidance about life based on the practical historical reality stemming from everyday problems and issues.'[29] Wisdom 'would encourage caution on the part of humans rather than impulsive action'.[30] Caution and prudence are called for, not prohibition.

Turning to the Protestants, American evangelical bioethicist, C. Ben Mitchel, sets the issue up this way: 'I want to reserve healing for restoration of lost capacities, and enhancement for improving on the species, or on the statistical norm.'[31] Interestingly, we find within American evangelical circles attempts to connect the blurry line between therapy and enhancement with the fall in the Garden of Eden. Would medical interventions that seek to restore what was lost in the fall count as therapy? Would the lengthening of life or even the overcoming of death count as therapy; or would it count as enhancement? This is not easy to sort out theologically.

> If, for instance, one holds that physical death and aging are a result of the Fall, then it would appear that a genetic battle against aging and death could be

---

26 Mathuna, 'Genetic Technology', p. 295.

27 James J. Walter, 'Catholic Reflections on the Human Genome', *The National Catholic Bioethics Quarterly* 3.2 (Summer 2003), pp. 275–86 (281).

28 Deane-Drummond, *Brave New World?*, p. 29.

29 Celia E. Deane-Drummond, *Creation Through Wisdom: Theology and the New Biology* (Edinburgh: T. & T. Clark, 2000), p. 234.

30 Celia E. Deane-Drummond, *Biology and Theology Today* (London: SCM Press, 2001), p. 117.

31 C. Ben Mitchel, 'Define "Better"', *Christianity Today* 48.1 (January 2004), p. 43.

justified, and therefore qualify as therapy... However, more recent shifts in theological thinking, not immune to the profound influence of evolutionary theory
where both aging and death are an *integral* part of how we humans have come to
be, have resulted in a rejection of the idea that physical death was introduced into
the world by Adam's sin.[32]

In the latter case, life extension through genetic engineering would count as
enhancement.

Several centrist Protestant ethicists note how difficult it is to draw the line
between therapy and enhancement. Yet, some ethicists still believe it is possible
to specify a baseline of human functioning that is part of the intended order of
creation. When someone falls below this baseline, genetic interventions are called
therapy or correction. When an intervention would move someone above the baseline, it is called enhancement. Paul Ramsey has argued strenuously for an 'exact'
and limited meaning of the term 'genetic therapy', precisely so that it could not
be used for enhancements that treat only desires.[33] If we would intervene in our
own genetic code to improve on human nature, we would usurp what rightfully
belongs solely to God. We would be playing God. Ramsey is remembered for
having penned these lines in 1970: 'Men ought not to play God before they learn
to be men, and after they have learned to be men they will not play God.'[34] Such
an action would constitute human pride or *hubris*.[35]

Leroy Walters and Julie Gage Palmer have asserted that 'disease and disability'
are 'evils that should be overcome as quickly and efficiently as possible'.[36] While
all children of short stature may experience pain or discrimination, the question of whether there is a physiological basis for the short stature is taken to be
morally relevant for policy purposes. 'We are attempting to draw a sharp line
between *bona fide* illness...and physical traits that can lead to discouragement
or discrimination or both...'[37] They approve genetic enhancement for children of
short stature who have hormone deficiencies, but not for children of short stature
who do not have hormone deficiencies.

James C. Peterson is one of the most thoughtful Protestants to take up the issue
of enhancement. The line between therapy and enhancement is blurry, he admits.
'The problem is both with the indistinctness of the line and with the rationale for
holding it. There is no adequate conceptual distinction between cure of disease
and enhancement of capacity that would allow us to make a principled argument
for cure of disease that would not over time also allow genetic intervention for

32   Todd T. Daly, 'Therapy vs. Enhancement: The Problem Posed by Anti-Aging Technologies',
2005, The Center for Bioethics and Culture Network, www.thecbc.org/redesigned/research_
display.php?id=199.

33   Paul Ramsey, 'Genetic Therapy: A Theologian's Response', in Michael Hamilton (ed.), *The
New Genetics and the Future of Man* (Grand Rapids, MI: Eerdmans, 1972).

34   Paul Ramsey, *Fabricated Man: The Ethics of Genetic Control* (New Haven: Yale University
Press, 1970), p. 55.

35   For an analysis of the concept of 'playing God', see Ted Peters, *Playing God? Genetic
Determinism and Human Freedom* (London and New York: Routledge, 2nd edn, 2002).

36   Leroy Walters and Julie Gage Palmer, *The Ethics of Human Gene Therapy* (Oxford and New
York: Oxford University Press, 1997), p. 113.

37   Ibid., p. 113.

enhancement.'[38] Peterson works with an anthropology and ethic of creative love. 'If human beings are called to develop themselves, purposeful and direct enhancement of capacity could be appropriate, or according to some even required.'[39]

5. *Transhumanism and the Question of Embodied Identity*

All the above theological discussion pertains to the human body. We find theologians making two assumptions regarding our identity. First, who we are as a person is embodied. Second, changes in the body, even if resulting in changes in the mind, do not risk a loss of identity. Beyond the therapy and even beyond the enhancement, our transformed self will still be our self. When we travel the path from therapy and enhancement toward transhumanism, however, these two assumptions do not travel with us. Theologians will eventually need to consider the implications of neuroscience aiming at transhumanism.

Curiously, the assumption at work in transhumanism is that human intelligence and hence human personhood can become disembodied. This is by no means a return to substance dualism and belief in a metaphysical or spiritual entity; yet, it is still curiously akin to pre-modern beliefs in disembodied intelligent existence. Information patterns have simply replaced the disembodied soul.

Old-fashioned body lovers object. 'Modern transhumanism is a statement of disappointment', writes Brian Alexander. 'Transhumans regard our bodies as sadly inadequate, limited by our physiognomy, which restricts our brain power, our strength and, worst of all, our life span. Transcendence will not be found in the murky afterlife of the usual religions, but in technological and biological improvement.' [40]

Objections come from at least two allied directions, from science and from theology. Scientifically, we are becoming increasingly convinced that brains and hence minds are embodied, perhaps even communal. Despite the significant role played by our genetic inheritance, we are learning that during our childhood years our brain functioning is itself developing; environmental factors have a decisive impact on brain development as well as formation of the human self. Brains, bodies and the environment are in constant interaction. The transhuman 'proposals are based on the confidence that the human mind can be explained wholly by a computational functionalist approach', is the assessment of Gregory Peterson. 'While there are important reasons to consider such a possibility, it is by no means certain... Is a downloaded version of me the same me? ... We have become increasingly aware that the mind is intimately connected to physical states. We are not simply disembodied reasoning machines but persons in a bodily and communal context.'[41]

---

38   James C. Peterson, *Genetic Turning Points: The Ethics of Human Genetic Intervention* (Grand Rapids, MI: Eerdmans, 2001), p. 252.

39   Ibid., p. 288.

40   Brian Alexander, *Rapture: How Biotech Became the New Religion* (New York: Basic Books, 2003), p. 51.

41   Gregory R. Peterson, *Minding God: Theology and the Cognitive Sciences* (Minneapolis: Fortress Press, 2003), p. 218.

Christian theology during the modern period has reaffirmed the goodness of the body and jettisoned the substance dualism of the past. 'Transhumanism is in some ways a new incarnation of Gnosticism,' complains Mayo Clinic hematologist and bioethicist, C. Christopher Hook. 'It sees the body as simply the first prosthesis we all learn to manipulate.'[42] Hook goes on to declare that 'As Christians, we have long rejected the Gnostic claims that the human body is evil. Embodiment is fundamental to our identity, designed by God, and sanctified by the Incarnation and bodily resurrection of our Lord.'[43] This theological affirmation of embodied human existence suggests unqualified commitment to therapy, perhaps commitment to enhancement, and scepticism about transhumanism.

What if a transhumanist would provide us with a new body, a simulated replica of our present body for the use of our pattern of information processing? Could we live forever? Would this meet the theological concern for embodiment?

Key to opening the transhuman door to the future is the development of Artificial Intelligence (AI). The link between AI and the transhuman vision is a series of assumptions: (1) what makes us human is our intelligence; (2) intelligence consists of information processing; and (3) the transfer of the pattern for information processing from our brain to a machine is feasible. Further, if we keep the machines in perpetual repair, we can live forever. We will have achieved cybernetic immortality.

Such a promise tantalizes. Tulane physicist Frank Tipler combines AI and cosmology to project an image of a future state of intelligent existence that constitutes immortality. We will create computers in our image, and these computers in turn will create a replica of ourselves; and this replica will persist beyond our death and beyond the death of the universe.

Tipler's method unfolds the religious candy wrapper while setting aside the religious candy inside. Even though he mentions the Easter resurrection of Jesus and the eschatological writings of Paul, he tries to promulgate a strictly secular immortality for techno-sapiens.[44] Tipler bases his promise on the history of evolutionary progress. Resurrection here will be the result of a future evolutionary event in which life understood as information processing will take hold of its own destiny and create a supra-biological environment for its existence just prior to the moment when the physical world self-destructs.[45]

Tipler addresses a couple of concerns theologians would likely raise here: perfection and identity. First, he is sensitive to the human yearning for perfection. Our present state of existence is not satisfactory. We do not hunger simply for life beyond death. We hunger for salvation. So, without using the term 'salvation', Tipler announces that the perpetuation of our pattern of information processing

---

42 C. Christopher Hook, 'The Techno Sapiens are Coming', *Christianity Today* 48.1 (January 2004), p. 38.

43 Ibid.

44 Frank J. Tipler, *The Physics of Immortality* (New York: Doubleday, 1994), pp. 305; 309–313.

45 Tipler, *Physics of Immortality*, p. 225.

within our simulated body will transcend the previous model by eliminating bodily defects such as missing limbs; youth will be substituted for old age; sight for blindness; and so on.

The second concern is whether continuity of identity will be maintained. Tipler answers, yes. Anticipating objections that total death followed by total re-creation denies continuity, Tipler responsively argues that continuity in conscious self-identity is both necessary and possible. Replication is not annihilation. To be resurrected as a replica of one's former self does not deny that it is the same self. The identity of the information patterns within which we are aware of our experience of the world and ourselves seems to be sufficient for Tipler. In sum, what is resurrected is the immaterial form but not the material substance of who we presently are.

> An exact replica of ourselves is being simulated in the computer minds of the far future. This simulation of people who are long dead is 'resurrection' only if we adopt what philosophers call the 'pattern identity theory'; that is, the essence of identity of two entities which exist at different times lies in the (sufficiently close) identity of their patterns. Physical continuity is irrelevant.[46]

Is this pattern theory merely another example of platonic body–soul dualism in which a nonmaterial soul is extracted permanently from a material base? No, says Tipler. Tipler's simulation so emulates the physical body that, for all practical purposes, what resurrected souls experience is physically real. According to Tipler, resurrected souls experience themselves in their environment; and this environment is experienced as if it were physical. In a surprising move, Tipler reiterates Bishop William Berkeley's subjective idealism: to be is to be perceived.[47] If as a computer simulation we perceive physicality, the physicality exists thereby.

Let me ask a theological question: is the replication of a pattern of information processing sufficient? Let me be sympathetic for just a moment. Rather than requiring God to locate and piece together all the molecules of our previous body, could God's task at the resurrection be simplified by merely reassembling the pattern of our body's molecules? Rather than the molecules as matter, might we say that the form is what counts? Could God get by with merely remembering and reincarnating our form or pattern? Tipler is reminiscent of Origen: 'The previous form does not disappear, even if its transition to the more glorious occurs…although the form is saved, we are going to put away nearly [every] earthly quality in the resurrection…[for] "flesh and blood cannot inherit the kingdom" (1 Cor. 15:50).'[48] For Origin and Tipler the form is saved, but not the substance. Is this sufficient to guarantee continuity of personal identity? I doubt it. Our identity is the product of our biography. Our identity is something gained over time, inclusive of scars embedded in our limbs combined with our remembered

---

46 Tipler, *Physics of Immortality*, p. 227.
47 Tipler, *Physics of Immortality*, p. 211.
48 Origen, 'Fragment on Psalm 1:5', cited by Carolyn Walker Bynum, *The Resurrection of the Body in Western Christianity*, 200–1336 (New York: Columbia University Press, 1995), p. 64. My original analysis of Tipler can be found in Ted Peters, 'Resurrection: The Conceptual Challenge', in Ted Peters, Robert John Russell and Michael Welker (eds.), *Resurrection: Theological and Scientific Assessments* (Grand Rapids, MI: Eerdmans, 2002), pp. 297–321.

thoughts. Is there not more to our individual consciousness than information processing, namely, our own information processing informed and influenced by the accumulated history of our bodily functions?

Let me ask an additional question: if the key to cybernetic immortality is replication of the soul's pattern, what would happen in the event of multiple replications? We already know what computer clones are. In principle, the soul's pattern could be replicated many times, not just once. Which pattern would maintain the individual's identity? Does continuity of unique personal identity require some degree of substance uniqueness?

Just how realistic is the Tipler theory, scientifically speaking? Not very, either physically or biologically. John Polkinghorne dubs these ideas as

> excessively speculative in the assumptions that they make about physical processes in unexplored circumstances, particularly in Tipler's case. The closing instants of a collapsing universe involve physical processes at energies vastly in excess of those of any regime of which we could claim to have an understanding... The speculations of the physical eschatologists are also chillingly reductionist in tone. Life is equated to the mere processing of information. Only if one believes that humans are no more than computers made of meat could one regard their replacement by computers made of bizarre states of matter as affording a picture of continuing fulfillment.[49]

Just how realistic is cybernetic immortality, philosophically speaking? Although the late philosopher Paul Ricoeur does not address Tipler directly, he says something quite relevant. The very concept of the brain and the very work of neuroscience are products of the human mind. This means the mind has sufficient independence so as to authorize its own self-study, something the brain certainly could not do on its own. 'I propose we adopt the term *substrate* to denote the relation of the body-as-object to the body as it is experienced, and therefore of the brain to the mental.'[50] By *substrate* Ricoeur means something like Aristotle's material cause. Ricoeur makes this move to challenge the unfounded assumptions of the scientists that the brain is the efficient cause of human mental operations. Neuroscientists over-extend themselves, complains Ricoeur. At best, we can identify a correlation between brain activity and mental activity; but no warrant exists for a causative relationship. 'The brain is the substrate of thought (in the broadest sense of the term) and...thought is the indication of an underlying neuronal structure. Substrate and indication would thus constitute the two aspects of a dual relation, or correlation.'[51] Ricoeur puts a fence up against reducing human consciousness and mental activities to the pattern of neuronal functioning. The implication is that even if the pattern of brain function could be replicated, there is no guarantee – not even a likelihood – that the consciousness of the person would be replicated with it.

49   John Polkinghorne and Michael Welker (eds.), *The End of the World and the Ends of God* (Harrisburg, PA: Trinity Press International, 2000), p. 33.

50   Jean-Pierre Changeux and Paul Ricoeur, *What Makes Us Think?* Trans. M.B. DeBevoise (Princeton, NJ: Princteon University Press, 2000), p. 46.

51   Changeux and Ricoeur, *What Makes Us Think?*, p. 47.

Just how realistic is cybernetic immortality, theologically speaking? Noreen Herzfeld is critical. 'The assumptions regarding the natures of the human person and of eternal life that underlie the hope of an immortal presence within computers are quite different from those of most Christians. Cybernetic immortality assumes a dualistic understanding of the human person, a conception of eternity as "a long time", and a hubristic faith in human power.'[52] Cybernetic immortality cannot become a substitute for what Christians understand as resurrection of the body.

What are the ethical implications? Denigration of the body is one implication. Herzfeld continues:

> Our finite bodies are an integral part of who we are. The essential nature of the human being always contains two inseparable elements, self-transcending mind and finite creaturely being. The denial of the latter has led to a denigration of both women and the natural environment. Cybernetic immortality leads directly into these twin denigrations… It is notable that cybernetic immortality had been suggested as a possibility only in the writings of rich, white males.[53]

The cybernetic immortality projected by Frank Tipler represents transhumanism in its most audacious and extreme form. Much more modest forms of integrating human brains and computational machines could fit within the broader categories of therapy or, more likely, enhancement. Instead of thinking of cyborg brains as transhuman, we could think of them simply as fabulously human.

6. Eschatological Transformation and Perfected Human Beings

'Render therefore unto Caesar the things which are Caesar's; and unto God the things that are God's', says Jesus (Matthew 22.21, KJV). In our case, Caesar is science. Any ethic should begin by distinguishing realistically what belongs in the domain of science from what belongs in the domain of divine promise. Once this distinction has been made, then we can look for the connection. God has promised an eschatological transformation, the advent of the new creation. Our task this side of the new creation is to engage in the much more modest work of transformation in order to improve human health and flourishing.

The eschatological orientation seeks the good in the future, not the past. It presumes the world anticipates its own betterment, that our human condition yearns for redemption. Human enterprises such as scientific research and medical advance contribute to healing and overall human wellbeing. They are attuned to the work of our God of transformation who promises to make all things new. Under the right conditions, science can be considered a godly vocation.

I like to speak of an eschatologically oriented ethic as a 'proleptic ethic', with 'prolepsis' indicating that today's actions anticipate tomorrow's transformation.[54] A proleptic ethic begins with a vision, a vision of the perfected human being

---

52 Herzfeld, *In Our Image*, p. 73.
53 Herzfeld, *In Our Image*, p. 74.
54 Ted Peters, *GOD – The World's Future: Systematic Theology for a New Era* (Minneapolis: Fortress Press, 2nd edn., 2000), chapter 12.

residing in a new creation. 'And God shall wipe away all tears from their eyes; and there shall be no more death, neither sorrow, nor crying, neither shall there be any more pain: for the former things are passed away' (Revelation 21.4, KJV). The Bible's closing apocalypse envisions a total healing. Scientific and medical contributions to human betterment today, modest as they may be, anticipate, and thereby participate in, God's final redemption.

The task of the ethicist, I believe, is to devise middle axioms that connect the grand eschatological vision of a new creation with our quite human responsibilities in the present time. Middle axioms would be theological principles providing ethical support for scientific research and medical technology that contribute to improved therapy and perhaps some degree of enhancement. Theologically, we need to affirm that scientific and technological transformation actually participate in the renewing of our world in a way that is both human and divine. Yet, blanket baptism of anything and everything new would be imprudent, to say the least. Constructing such middle axioms requires wisdom and prudence, what the Aristotelian tradition knows as *phronesis* and hermeneutical philosophers know as *applicatio*. Such application incorporates the unavoidable ambiguity inherent in assessing the practical outcomes of speculative proposals. The task of the ethicist remains: face the ambiguities, invoke wisdom, think prudently, and render the best judgement that finite considerations can produce.

7. Enhancement and Justice

As we turn to our three areas of brain research – therapy, enhancement and transhumanism – my own considered judgement is the following. Therapy or healing is incontrovertibly a divinely appointed ministry as well as a humane enterprise. Transhumanism can be dismissed because it is scientifically and philosophically unrealistic as well as theologically and ethically misdirected. When it comes to enhancement, however, we confront complexities and nuances that may require additional analysis.

When we ponder the standard definition of 'enhancement', we must take note of the competition component. By definition, enhancement seems to require that one individual become superior to others, that one person stands out as being above average. Is enhancement necessarily anti-egalitarian? Could widespread enhancement technologies lead to a new eugenics movement, in which some segments of society benefit and others suffer discrimination? If so, then we must face the justice issue.

In the United States, the National Council of Churches has raised cautions about eugenic programmes.[55] Similarly, the United Church of Christ in 1989 welcomed the development of genetic engineering *provided* there would be appropriate regulation and 'justice in distribution'.[56] The United Church of Canada expressed as a general principle that the 'rights of the weaker and the needy' must be protected

---

55   Ronald Cole-Turner, *The New Genesis: Theology and the Genetic Revolution* (Louisville, KY: Westminster/John Knox Press, 1993), p. 73.

56   Cole-Turner, *New Genesis*, p. 76.

in any genetic interventions.[57] Karen Lebacqz has also warned that expensive advances in medical genetics may be 'no deal for the poor'.[58]

These observations regarding justice uncover the heart of the problem with enhancement in all its forms. If enhancement comes to mean that the capacity of one individual is raised to a level superior to others, the question of justice arises. For the near and medium range future, medical technologies will be expensive. Only some segments of society will be able to afford such medical advances. The financially rich may also become the 'genrich', to use the language of Lee Silver.[59]

'Justice is concerned broadly with the idea that each is given her or his due,' writes Deane-Drummond. 'Unlike many other virtues, justice specifically governs relationships with others.'[60] The aspect of enhancement which draws our ethical scrutiny here is its relationship with others. The risk or threat is not found in improving one's memory or executive capabilities; the justice problem arises due to the relative social status of such a person once this has been accomplished. The threat of injustice is found in the likelihood that privileges will go to the enhanced; and the unenhanced will suffer marginalization.

Professional sports are currently rocked with scandal due to drug-induced physical enhancements.[61] Intuitively, the public knows this is unfair, unjust. We can expect to see legal rules put in place to keep the competition egalitarian. Might such policies be devised to govern genetic and neurocognitive enhancements? Enhancement technology could be parsed accordingly. Does enhancement necessarily lead to social inequality?

Now, we might reverse the logic for a moment. Suppose enhancement for some and not others does in fact lead to social inequality. Let us pose some other justice questions. Should we deny to an individual the opportunity for neuro-cognitive enhancement simply because others in the community might not have equal opportunity? Should the fear of social or economic injustice be sufficient to warrant denial of increased wellbeing for an individual? Such questions require attention by this generation of theologically guided ethicists.

In light of this, would it be prudent to alter our understanding of 'enhancement' and, thereby, our understanding of the perfect human being? Could we eliminate the social competition component? Rather than work with a standard baseline or human average with regard to memory or executive abilities, could we simply ask what enhancement could mean on a case-by-case basis? Could we think in terms of incremental improvements in individual wellbeing? That is to say, could

---

57   Audrey Chapman, *Unprecedented Choices* (Minneapolis: Fortress Press, 1999), p. 60.

58   Karen Lebacqz, 'Fair Shares: Is the Genome Project Just?' and 'Genetic Privacy: No Deal for the Poor', in Ted Peters (ed.), *Genetics: Issues of Social Justice* (Cleveland: Pilgrim Press, 1997), pp. 239–54 (239).

59   See L.M. Silver, *Remaking Eden: How Genetic Engineering and Cloning Will Transform the American Family* (New York: Avon, 1998).

60   Deane-Drummond, *Ethics of Nature*, p. 15.

61   American baseball player Jose Canseco describes himself as 'a nearly superhuman athlete' due to his use of steroids and growth hormone. *Juiced* (San Francisco: HarperCollins, 2005), p. 2.

we devise an ethical principle that would exclude enhancement for competitive purposes but encourage it for individual flourishing? Again, these are questions deserving of attention.

8. *Conclusion*

In summary, here we have followed the track of therapy, enhancement and trans-humanism en route to neurocognitive improvement through genetic engineering, neurotechnology and pharmacology. We have assessed the direction this is going in light of a proleptic ethic buttressed by a virtue ethic that emphasizes wisdom and prudence. We have found that the concept of therapy would welcome any advances in neuroscience that could lead to healing of brain and mind. We have found that transhumanism is founded on unsubstantiated assumptions regarding the human person and human intelligence; and that theologically transhumanism confuses ultimate transformation with the more modest possibilities of genuine improvement in human wellbeing.

Finally, we have followed the track of the enhancement train. With regard to the individual, enhancement of mood, memory, or executive capabilities in them-selves could lead to an improvement in human wellbeing without raising ethical objections. However, ethical problems arise when enhancement is used to create a fabulous human being who is superior to others in the community and who benefits unjustly from this superiority.

It is my own considered judgement that the risk of injustice is insufficient to slow down or curtail laboratory research on improving neurocognitive capacities. The task of the ethicist should be applied to distribution of the medical benefits so as to meet the standards of social and economic justice.

Chapter 2

Human Nature and the Human Condition

Gordon Graham

1. *Crossing Ethical Boundaries*

There is a widely held belief that the advent of sophisticated genetic technologies has opened the door on a new and daunting range of ethical problems, and that this is because genetic technology makes it possible for human beings to do things that they could never have contemplated doing before – cloning and genetic modification, for example. If this is true, it is not a feature unique to genetic technology. Nuclear technology may be said to have done the same, for, though the danger of their use has receded with the end of the Cold War, it remains the case that the invention of nuclear weapons gave human beings, for the first time in history, the power to destroy the entire planet on which they live.

There is, however, a difference between the two cases. We can identify easily the awful prospect nuclear war presents, and say precisely where its ethical dimension lies. The traditional Just War proscription against killing non-combatants rests on the moral principle that the innocent ought not to be killed. The nature of nuclear weapons is such that their use would result in the death of vast numbers of non-combatants. Thus, though nuclear weapons are new, the ethical principles applicable to their use are the same as they have always been.[1] It is sometimes suggested that the moral issues surrounding nuclear technology are not confined to weapons of mass destruction, but involve the use of nuclear power quite generally. The technology of nuclear fission has opened up the prospect of our unleashing physical forces that we cannot then contain, a modern scientific equivalent of Pandora's Box. This perception is not the only factor that has brought about a reduction in the nuclear generation of power, but it has figured prominently among objections to it, despite the fact that concern over global warming and greenhouse gases might have been expected to advance the case for a form of energy that is free from carbon emission, as nuclear energy is.

Closer analysis of just what it is about this feature of nuclear power that might morally constrain our use of it usually reveals an appeal to what is known as 'the precautionary principle'. Accounts of this principle vary, and some interpret it as meaning little more than urging caution in assessing anticipated benefit over predicted cost. But a more robust interpretation makes important use of a contrast between the merely harmful and the catastrophic. In deciding what to do we

---

1 G. Graham, *Ethics and International Relations* (Oxford: Blackwell, 1997).

ought to avoid harmful outcomes, but the nature of life is such that it is not in fact always possible to do so. Consequently, if human beings are to act at all, they have to run the risk of their actions, however well intentioned, proving harmful to themselves and others. Accordingly, as an ancient watchword has it, probability must be the guide to life. The form this guide takes is easily stated: we should assess possible actions in terms of the 'value of outcome' (V) multiplied by 'probability of its coming about' (P). This means that we are justified in embarking on potentially harmful courses of action if there is a reasonably low probability that their undesirable consequences will come about, and a higher probability that their advantageous outcomes will be realized.

Against this background, the precautionary principle states that, where there is the prospect not merely of harm, but of *catastrophe*, then V has such a high negative value that we cannot be justified in proceeding with the action in question, *however* low P may be. In short, by combining the desirability of avoiding catastrophe with a probabilistic account of practical reason, the precautionary principle generates an absolute ban on certain actions, and by extension, certain government policies.

The precautionary principle has been invoked by opponents of genetic modification and research on human embryos, as well as the opponents of nuclear power. The problem is, however, that its serious application leads to more than precaution; it leads to paralysis. Everything turns, of course, on the definition of 'catastrophe', but if this means 'extremely harmful outcomes', then there are very few courses of action (if any) that have no probability whatever of leading to such outcomes. The two world wars began with a series of initially relatively innocuous actions, not with acts of reckless endangerment. Yet these actions led to terrible outcomes. But if all (or almost all) actions have a minuscule probability of producing catastrophic outcomes, the precautionary principle prohibits us from doing anything.

More importantly, perhaps, the slogan 'nothing ventured, nothing gained' applies to large-scale outcomes as well as to small-scale ones. To place an absolute ban on actions that have even the smallest chance of ending in catastrophe is also to put a ban on actions that offer the prospect of immense benefits.[2] This feature of the precautionary principle is especially important when applied to genetic technology. To invoke it in support of a total ban on the genetic modification of organisms or research on human embryos may avoid potential catastrophe. But by the same token, it denies humankind the potentially vast benefits that these same technologies might produce.

In any case there is this further point. The precautionary principle is consequential. That is to say, it seeks to regulate and proscribe behaviour in accordance with anticipated consequences. By contrast, the widely held belief with which we began intuitively sees something *intrinsically* wrong with the actions that genetic technology has made possible. The thought is that if there is something deeply objectionable about genetic enhancement (for instance), this is not just a matter of

---

2 See N. Manson, 'The Precautionary Principle, the Catastrophe Argument and Pascal's Wager', *Ends and Means* 4.1 (1999), www.abdn.ac.uk/philosophy/endsandmeans/vol4no1/manson.shtml.

the consequences it might have, however good or bad. There is something wrong about it *in itself*.

It is this sense of intrinsic error that the language of 'playing God' signals. The idea can be spelt out theologically,[3] or in a quasi-theological way.[4] But it is also an expression that has been used even by authors who mean nothing properly theological by it.[5] This suggests that some further way in which the intrinsic error of certain genetic technologies can be articulated is required, an articulation that can encompass both the theological and the non-theological. One interesting idea in this connection is that the boundary genetic technology is in danger of crossing is the boundary that lies between the age-old endeavour of ameliorating the human condition, and the new ambition of changing human nature itself. However, this way of stating the issue begs an important question. Some recent literature on 'transhuman' and 'posthuman' futures does not regard such a boundary crossing as a danger, but an exciting new opportunity. The purpose of this essay, then, is to ask first, whether the distinction between human nature and the human condition does indeed constitute a significant boundary relevant to technology, and second, whether crossing it is a prospect properly regarded with fear or enthusiasm.

2. *Human Nature*

The concept of human nature has ancient roots. Both Plato and Aristotle make extensive use of it, though not always by that name; the Greek '*telos*' is usually translated 'purpose' or 'end', but is employed to delineate an intrinsic nature also. In the modern period, however, the concept of human nature has had a more chequered intellectual history. In the 18th century it formed the basis of a great and innovative intellectual project – the 'science of man' – and expressly features in the title of this project's most famous work – David Hume's *Treatise of Human Nature* (1739/40). Hume's *Treatise* aimed to set out, on the basis of empirical observation, the universal features of human perception, mind and emotion, and the law-like principles on which they operate, chief among them the 'association of ideas'. It was an ambition shared by many other thinkers, notably the philosophers of the Scottish Enlightenment – Francis Hutcheson, Adam Smith and Thomas Reid especially. But by the middle of the 19th century, enthusiasm for this project faded, partly because the new 'science' had made much less progress than expected, and partly because philosophers became more conscious of the historical particularity of features that had formerly been taken to be universal endowments of the human mind. In particular, the moral and aesthetic sense that most philosophers of the Scottish Enlightenment regarded as a central feature of human psychology, turned out to be almost as variable across cultures as the

---

3   See T. Peters, *Playing God?: Genetic Determinism and Human Freedom* (London and New York: Routledge, 1997).

4   See G. Graham, *Genes: A Philosophical Inquiry* (London and New York: Routledge, 2002), chapter. 4.

5   See J. Goodfield, *Playing God: Genetic Engineering and the Manipulation of Life* (New York: Harper & Row, 1977).

sense of humour. Secondly, an important part of the evidence that the science of human nature called upon lay in linguistic structures. Philosophers and linguists subsequently came to see that natural languages are important ways in which the mind conceives reality, and of course natural languages are historically particular and can be radically different.

In the late 19th and early 20th centuries, the rise, prominence and popularity of social anthropology both confirmed and emphasized this particularity, to the point where the social construction of reality, including the reality of humanity itself, came to be taken as an incontestable truth. In alliance with Marxists (with the occasional exception), anthropology made the concept of human nature positively suspect, a pseudo-scientific device by which the assumptions and values of one culture – that of Western Europe broadly understood – was imposed as the standard by which other less familiar cultural constructions were to be judged. 'Human' nature was in fact the nature of European man, dressed up as universally valid. Freudianism, it is true, ran against this general current, for it analysed the mind in terms of a universal threefold structure – ego, super-ego and id – but after a short period of extraordinary influence, its intellectual credentials faded.

Then, as a result of advances in evolutionary biology and sociology, the position changed again. With a combination of the two – 'sociobiology' (the term invented by E.O. Wilson[6]) – it seemed plausible to delineate enough recurrent and abiding features of human mind and action to speak once again of a universal 'human nature', a term that reappears in the title of one of Wilson's books.[7] The revolution in thought that made this possible was the advent of genetics. The discovery of the genetic basis of biological life combined with the insights of Darwinian evolution offers an understanding of the differentiating nature of species which, like the old science of man, is firmly based in empirical observation, but at the same time holds out the prospect of accommodating both change and variability. In this way the concept of human nature has come to be refurbished along relatively unobjectionable lines, a concept no more socially or historically value laden than the concept of 'cat nature' or 'fern nature'. Not all cats are the same. Nevertheless, they have a common genome and a shared evolutionary history.

It is the human genome that underlies this new conception of human nature (though the fact that genes are shared across species makes the position a complex one). The two are not identical, however, because as it has evolved the human genome (like that of other species) generates and sustains recurrent physical and psychological patterns that are the subject matter of ethology and evolutionary psychology rather than genetics. 'Human nature' is a complex of the two, but the genome, and our knowledge of it, is especially important since it gives us technological access to the structures that underlie our nature, and thus the theoretical possibility of changing it.

---

6   See E.O. Wilson, *Sociobiology: The New Synthesis* (Cambridge, MA: Harvard University Press, 1975).
7   See E.O. Wilson, *On Human Nature* (London: Penguin Books, 1995).

3. *The Human Condition*

A related concept that has received relatively little philosophical or scientific attention is that of 'the human condition'. That is not to say that such a conception is unfamiliar. 'The human condition' is often the subject matter of artists and writers. It is in fact the title of a painting by the Surrealist Rene Magritte. Nor is it absent in the investigations of the 18th century. Hume and Smith, for instance, write at some length about the conditions of scarcity within which human beings operate. Hume expressly distinguishes between 'natural' and 'artificial' virtues. The former – such as pride and sympathy – are those characteristics that are part of our nature. The latter – justice being the chief of them – derive from the fact that the world is such that the things we need to survive are in limited supply. Just as (say) the tendency to favour kin relations is a feature of human nature, so the relative scarcity of food, shelter, and so on, is a fact of the human condition.

Surprisingly, those who employ the modern scientific version of human nature rarely mention such a thing, even though the human condition is just as crucial a part of evolutionary explanation. This is because evolution makes central use of the idea of survival – in Darwin's version the survival of the fittest organism by natural selection,[8] in Dawkins's version the survival of 'the selfish gene'.[9] But *what* it is that organisms or genes have to survive is external to their nature. The general point applies to evolved humanity. Human beings are susceptible to certain diseases, for instance. For this susceptibility to be a challenge to survival, there have to be diseases of that kind; there are diseases to which human beings are *not* susceptible. But the existence of those diseases is not part of human nature; it is part of the human condition. Similarly, human beings are vulnerable to injury and death. It is this vulnerability that explains the predominance of certain aspects of their nature – the accuracy of their sight and hearing, for example, and the ability of their bodies to heal.

Hume and Smith (following earlier philosophers such as Locke and Hobbes) identified certain practices and social institutions as the central ways in which human beings had characteristically attempted to ameliorate the human condition, and with considerable success. Chief among these are the division of labour, the formation of the State and the creation of markets. Material goods are more plentiful if the labour involved in their production is divided in a way that favours specialization. Both goods and services are made more widely available through the bargains and exchanges characteristic of markets. Competing claims to limited supplies are settled with less harm and danger if there are laws to regulate them.

The perpetual effort to ameliorate the human condition explains a huge number of developments in cultural history. The shift from the life of nomadic hunter to that of settled agrarian is to be understood as the attempt to provide more adequately for basic needs by producing food and shelter rather than simply searching for

---

8 See C. Darwin, *The Origin of Species* (London: John Murray, 6th edn, 1872).
9 See R. Dawkins, *The Selfish Gene* (Oxford: Oxford University Press, rev. edn, 1989 [1975]).

it. The same effort also explains the creation and development of technologies. What we call the agricultural and industrial revolutions had immense social and political repercussions, but they are chiefly significant for ushering in a profound change in the ability of human beings to ameliorate the human condition, first and foremost by making available unprecedented quantities of goods and thereby combating natural scarcity to an extent hitherto unimaginable.

The development of better technologies is enhanced (to some extent) by greater knowledge. Though the dependence of technology on science is easily misunderstood and often exaggerated, there is enough of a connection between the two for it to be indisputable that science and technology together have played an important part in efforts to improve human life by ameliorating the human condition. Part of that improvement, of course, is to be found in medicine. Once more there is the possibility of overestimating the role of knowledge here. It is certain that better sanitation has contributed far more to human health than most advances in medical science; at the same time it is equally certain that modern medicine has increased our ability to combat disease, injury and ill health immensely.

There is no reason why this should not continue, and no reason why the science of genetics and the technologies it generates should not play their part in this. To the extent that they do, they are simply another phase in a very long history. In the course of that history there have been moral mistakes and economic disasters, many of them caused by ignorance and/or ambition. There is little doubt that similar failings will be found in the study and application of genetics. But this does not lend them any special moral or ethical novelty. On the contrary, it is precisely insofar as they simply follow the course of this long history that the moral questions they raise can be expected to be thoroughly familiar – how to secure benefit over harm, how to promote the general good without overriding individual right, and so on.

4. *Stepping Over the Line*

The widely held anxiety about genetic technology with which we began throws this last assertion in doubt, and holds that this new technology brings with it *new* moral dangers. The distinction between human nature and the human condition enables us to articulate this contention more clearly. Insofar as genetic technology seeks new and better ways by which to ameliorate the human condition, it is no more objectionable than the technology underlying computers or mobile phones. Almost anything that human beings discover or invent can be misused, of course. This is true of genetics, but not uniquely so. However, the position is quite different when the genetic technologist's ambition shifts from ameliorating the human condition to altering human nature itself. This is precisely the ambition that those who speak of a 'transhuman' or 'posthuman' world entertain.

To state the matter in this way is helpful primarily because it allows us to frame two central questions. Is there really a crucial line between human nature and the human condition? If there is, what exactly is wrong with crossing it? For reasons that will eventually become apparent, I shall consider these questions in reverse order.

Crossing the line could be an error in one of two possible ways. Considered at the most general level, either it arises from a serious moral failing, or it constitutes a monumental piece of folly. The two can be connected, of course, as in the classic sin of *hubris*. Prometheans, who set themselves against the gods (or God), are both guilty of gross impiety *and* immensely foolish. It is a topic to which we will return, but for the moment it is useful to hold morality and irrationality apart.

If the attempt to alter human nature is a moral failing of some sort, where does its wrongness lie? An earlier part of the discussion revealed that we should not look to the consequences of action, however grave, to provide a criterion of differentiation. Of course, careful and critical cost/benefit analysis is often relevant to practical deliberation in this context, but it cannot serve to discriminate between classes of action. It is the act itself that must be shown to be wrong. What this implies is that any attempt to alter human nature must violate some fundamental law. But what law? Human or positive law is evidently insufficient. Any campaign to make research on human embryos illegal is necessarily based upon the assumption that it is not already contrary to the law. The law in question must thus be of some 'higher' kind – the moral law or the law of God.

The problem with appealing to the law of God is twofold. First, just what the law of God proscribes is a matter of deep contention. This is evident both within and across religions. There are important differences between Muslims, Jews and Christians on what human conduct is and is not acceptable to God, and deep differences between the adherents of these same religions. In common with the vast majority of Muslims, for example, most Christians believe that homosexual conduct is contrary to the law of God. But not all do, and those who do not are equally convinced that God will bless faithful same-sex relationships. So is homosexuality against God's law or not? The difficulty in answering this question straightforwardly arises from the fact that what is and is not contrary to God's law is determined by Scripture, but even where there are Scriptures common across and between religions (the Old Testament and the Hebrew Bible, for example), they are open to interpretation. This fact, combined with their ancient origin, suffices to make their application to the novel circumstances created by contemporary culture and modern technology highly uncertain. In short, the attempt to alter human nature may indeed be contrary to the law of God, but it simply is not possible to show this with even a semblance of conclusiveness.

It is important to stress that this point should not be confused with a general religious scepticism. It is consistent to hold both that some actions cannot be shown to be contrary to God's law, and that there are also actions and attitudes which we need have no hesitation in condemning on this same ground; making God the subject of mockery, using consecrated places for mundane purposes, or systematically subverting liturgical offices (as in the Black Mass) are all clear examples of flouting the law of God. It is also possible to think of a less flagrant example more relevant to the present issue – failing to accord nature and 'the given' the respect and reverence which all religions hold it to warrant. The problem is to identify precisely those contexts and occasions when such respect and reverence can properly be said to be lacking. It is easy to imagine someone who takes the opposite sex for granted, seeing in them merely opportunities for

personal physical gratification. Contrast this with someone of a homosexual orientation who loves another person of the same sex and regards the relationship as a
wonderful gift for which God is to be thanked. In whom is reverence present and
in whom is it missing? The fact that this question is (at a minimum) hard to answer
indicates the uncertainty that can attach to even the most sincere attempt to apply
and live by God's law (an uncertainty that 'true' faith requires, perhaps).

But there is in any case this further difficulty. Even if the law of God could
be specified with sufficient precision to show that there is something wrong
with genetic engineering, this would not be persuasive to those for whom the
Scriptures on which this judgement was based possess no special authority, and
would present no obstacle to anyone for whom the concepts of God and 'the given'
mean nothing. Yet if the idea of crossing the boundary between ameliorating the
human condition and altering human nature is to have real purchase in the world
of genetic technology, it must have something to say to just such people.

Is there any greater mileage to be obtained from the idea that crossing this
boundary is a violation of a moral law? In the *Summa Theologica*, Thomas
Aquinas draws a distinction between four kinds of law – the eternal, the divine,
the natural and the human.[10] Leaving aside the first of these, this classification
creates space for a natural law that lies between the law of God (divine law) and
the law of the State (human law). Aquinas is here building upon Aristotle, and the
conception of natural law he invokes is strikingly Greek in character. The idea is
that each kind of thing has its proper functioning, a way in which it ought to go,
so to speak, and there is something wrong about any attempt to force it out of this
way. This conception of 'wrongness' is quite different to the Kantian conception
of wrongful action under the moral law, a conception that has been immensely
influential in European philosophy. Kant's moral law is the product of what he
calls 'pure practical reason', and it consists in demonstrable rules of conduct that
place every conceivable human action into one of three categories – the forbidden,
the obligatory, and the permissible. Though Kant's intellectual achievement was a
brilliant one, it is now widely agreed that he was ultimately unsuccessful and that
reason alone cannot demonstrate moral right and wrong. In its aftermath, and in
light of the limitations of the utilitarian alternative, philosophers have turned once
more to 'virtue theory', an approach to morality that takes its cue from Aristotle
and arrives at conclusions sympathetic to Aquinas's system of ideas. Yet, though it
has much to commend it, this way of thinking will not serve in the present context.
That is for two reasons. Many simpler and more familiar ways of going against
nature predate genetic technology, and these generate too many examples that fall
far short of anything we would want to describe as a moral wrong.

By illustration of this, consider the simple case of the seedless grape. The
engineering of a seedless grape (whether by genetic modification or artificial
selection over a number of generations) has resulted in something that could not
be found in nature and which, left to its own devices, could not survive in nature.
To manipulate the grape in this way is clearly to go against its nature. Nonetheless,

---

10   Thomas Aquinas, *Summa Theologiae*, XIV (London: Blackfriars in association with Eyre
and Spottiswoode, 1963), QQ 92–96.

it is difficult to say in all seriousness that those who first produced the seedless grape committed a grave moral error. It might be replied that the example is not a good one because it is local rather than global. Now 'local' might mean confined to one type of organism, which of course it is, or it might mean confined to a single generation. It is the second limitation that is more important because it has often been held that genetic modification is of far less concern if it is limited to a single individual than if it is one that will continue down subsequent generations. But it is hard to see that such a distinction does much moral work. It is certainly true that the seedless grape cannot propagate its seedlessness. Nevertheless, it can still be the case (and may already be, for all I know) that the seedless grape becomes the norm and that grapes capable of self-generation become extinct. In that case, the seedless grape does represent a global change.

Another line of thought is that the example of the seedless grape is too trivial to illuminate the more serious issues surrounding the manipulation of *human* nature. At most it is an oddity, not a monstrosity. What technologically contrived modification to the genome threatens is the unexpected emergence of monstrosities. Even if there is this difference in importance, however, two points are worth making. First, the prediction that meddling with the human genome will produce monstrosities is one that might not actually come true. Human beings are notoriously bad at prediction. Second, the sorts of changes that many genetic engineers have in view (whether possible or not) are not of this kind. The elimination of human beings with haemophilia or latent Huntington's chorea could hardly be described as 'monstrosities' even though they are in a fairly obvious sense 'alterations of nature'. And this raises another important dimension to the question. Nature itself throws up monstrosities – the anencephalic foetus, for example. More interestingly, this very possibility presents us with an alternative religious possibility, that nature should not be regarded as having the last word. Aquinas himself says that 'grace perfects nature' which implies that there is reason for human beings to look beyond the natural. But it may also be true that grace perfects the unnatural. John Merrick, 'the elephant man', was a natural monstrosity, but if the film made about him portrays his character correctly, then his extraordinary grace succeeded in lifting him far above the limitation of his natural endowments and gave him a beauty of a quite different kind.

To summarize: if 'stepping over the line' is construed as an intrinsic error of a moral or religious kind, it is very hard to spell out convincingly. Even if we lend the utmost seriousness to the idea of flouting the law of God, or adopt the most naturalistic ethical conception, it still seems impossible to identify clearly and incontestably the nature of this error. Consequently, the objection to crossing the boundary between ameliorating the human condition and changing human nature, if any, must lie in its being a sort of folly.

5. *Practicality and Rationality*

A familiar and longstanding conception of practical reason holds that it consists in adopting efficient means to desired ends. The objection to this way of thinking is that it leaves no space for the idea of irrational ends. Yet it seems plain that, as well as adopting wholly inadequate means to the ends we desire, we can set our sights on absurdly ambitious ends. I might decide to buy a motor car, and set aside a pound a month for the purposes of purchasing it. This is evidently a wholly inadequate means to a reasonable end. Alternatively, I might decide to buy every aeroplane in the world, and by dint of great personal economy set aside £1000 a week to this end. In this case, however, it is not just that the means are inadequate; the end itself is impossibly unrealistic, and it is this that makes my action irrational.

In this example, the impossibly unrealistic is evident, but it is not always so. There are goals that I can set myself whose unrealism it requires some reflection to accept. Suppose that I am a more than competent pianist, aged 45 or so. Were I to aspire to become a well-known concert soloist, this seems less absurd an ambition than (say) becoming a world-famous brain surgeon. But, in fact, it might be only a little less absurd, and to spend the remainder of my life trying to realize it could amount to wasting the time that has been given to me no less than if I had spent it trying to become a brain surgeon.

It is plausible to say something similar about the ambition to change human nature by means of genetic technology. It is not just that the means now available fall far short. Actually current technology is impressive and is still (one imagines) in relative infancy. Thanks to contemporary science, mapping the human genome was completed well ahead of expectations, and perhaps this is evidence that our knowledge is not merely growing but accelerating. Consequently, it is not ridiculous to think that the technology might soon be up to many of the tasks that it is frequently said will be commonplace in an imagined future. Rather, it is the goal this imagined future represents that is suspect, and what makes it suspect is the absurd ambition that it disguises.

The human genome, if evolutionary biology is to be believed, is the outcome of many millions of years of selection and adaptation. This process has made the existing genome hugely well adapted to the human condition, the circumstances in which human beings must not merely survive but thrive. One of its outcomes is the most complex entity in the known universe – the human brain. Now the ambition of re-fashioning this genome more effectively must rest upon the supposition that the accumulated results of a little less than 200 years of biological science will enable us to do better than indefinitely many years of evolution have done. What possible reason could we have to think this? In effect, the existing genome is the embodiment of vast amounts of information on how to survive and thrive over indefinitely many generations. The error at the heart of 'transhuman' dreams consists in thinking that our knowledge of the human genome is a knowledge of the information it embodies, and our ability to tinker with it an ability to make use of that information.

The absurdity of this ambition can be made even plainer (perhaps surprisingly) by pursuing a theological line of thought. At one time it was supposed that the contents of the earth were the outcome of the special creation of distinctive species by God. Then the theory of evolution, supported by very considerable amounts of empirical evidence, came along. This subverted the idea of special creation in favour of the evolution of differentiated species from a common stock. Across the world of science and learning, this theory is now so widely accepted that it seems incontrovertible. But if it is, where does this leave God? Not redundant, as many have supposed. Sometimes evolution by natural selection is opposed to 'intelligent design', but while it may be that there is more to the explanation of the world we see around us than natural selection, evolution and divine purpose are not inevitably in conflict. God may have chosen to give evolution by natural selection a considerable role in the process of fashioning the world. In other words, God in his wisdom has preferred to let things evolve rather than establish them from the outset by fiat and design. To take both God and evolution seriously, then, we have to suppose that evolution is the preferred path of supreme wisdom.

Now if this truly is the case, it throws the ambitions of 'trans-' and 'post-' humanists in a new light. Let our knowledge be as great as it can be, it will still not equal God's. Combined with the previous point, this tells us something important. It tells us that any human ambitions with respect to redesigning the human genome are inevitably suspect, and based on an absurd overestimation of the epistemic resources that the science of genetics has put at our disposal.

This way of describing the absurdity, of course, by setting human ambition in the context of God's omniscience, presupposes the existence of God. There are many participants in the debate about genetics who would deny this, and others to whom the very idea is meaningless. But the general point can still be made effectively to them. To repeat: the human genome as it has evolved embodies a vast amount of information, information accumulated over aeons of trial and error. How likely is it that a century of investigation on the part of human science has come up with results and established knowledge superior to that embodied in the product of this vast period of time? The reasonable answer seems to be 'not very likely'. It follows that any attempts at engineering new futures based upon the contrary supposition are acts of gross folly. If it is indeed true that God has preferred evolution over his own intelligent design, then such attempts are also acts of gross hubris. If not, then they remain acts of monumental folly.

6. *Conclusion*

It is now time to return to the question suspended in Section 4 – is there really a clear line between altering human nature and ameliorating the human condition? There are some simple examples that seem to bring it into doubt. Consider virulent diseases like smallpox. In tackling smallpox we could set about ameliorating the human condition by eliminating the bacillus, or change human nature by making everyone immune to it (which looks like what we have done). Does it matter, morally or in any other way, which strategy we choose? Actually, the example is not as clear as this way of stating it suggests. If immunity is secured by systematic and repeated inoculation rather than by genetic modification of the human genome, then we do not have a case of altering human nature in the relevant sense. And if what is imagined is indeed changing the human genome for good, then this is as fanciful as any other imagined future.

The fundamental point is this, however. Even if the application of the distinction is unclear in certain instances, it still allows us to articulate the following

sort of difference. Human beings are mortal, yet there is every point in attempting to reduce their vulnerability to shortages of food, contaminated water, life-threatening diseases, organic defects, accidents and injuries. It might even be the case that this attempt succeeds to the point where the vast majority of human beings can reasonably look forward to a full lifespan, just as it is the case in the developed world that infant mortality and death in childbirth are no longer the great scourges that they were. All of this, however, is to be contrasted sharply with an alternative idea, currently canvassed by more than one author[11] – that human life might be indefinitely extended by the genetic elimination of ageing, still more with the suggestion that we might so modify human nature as to make it immortal. The idea of a future in which the human condition is immeasurably transformed to the point where natural scarcities of food, shelter, medicines and so on are effectively ended for good, in which (to employ a familiar slogan) poverty is made history, is still a dream, but not a fanciful one. A world in which genetic engineering accomplishes what evolution has not is so fanciful that any attempt to realize it is likely to result in nightmare more than dream. But the nightmare lies, not in the monstrosities that will be created, but in the collateral costs that will inevitably accompany foolish projects.

---

11   See S.J. Olshansky and B.A. Carnes, *The Quest for Immortality: Science at the Frontiers of Aging* (New York and London: Norton, 2001) and S. Shostak, *Becoming Immortal: Combining Cloning and Stem-cell Therapy* (Albany, NY: SUNY Press, 2002).

Chapter 3

THE NATURE OF HUMAN WELFARE

Søren Holm

It is not our human shape or the details of our current human biology that define what is valuable about us, but rather our aspirations and ideals, our experiences, and the kinds of lives we lead. To a transhumanist, progress occurs when more people become more able to shape themselves, their lives, and the ways they relate to others, in accordance with their own deepest values. Transhumanists place a high value on autonomy: the ability and right of individuals to plan and choose their own lives. Some people may of course, for any number of reasons, choose to forgo the opportunity to use technology to improve themselves. Transhumanists seek to create a world in which autonomous individuals may choose to remain unenhanced or choose to be enhanced and in which these choices will be respected.[1]

1. Introduction

In modern bioethics it is often claimed that it is impossible to derive any normative conclusions from the fact that something is 'natural' and *a fortiori* that there is no such thing as 'human nature', and even if there was we could not derive any normative statements concerning what is good for human beings from human nature. Such arguments can be traced back to Hume's arguments in *A Treatise of Human Nature*[2] and have been very common in discussions of assisted reproduction. They are also frequently used in the current debates about the moral issues raised by using biotechnology for the purpose of human enhancement.[3] The first quote above from the World Transhumanist Association exemplifies the position. What is denied is not that humans have a biological nature that sets the species *Homo sapiens* apart from other species, but (1) that this biological nature has normative import and (2) that there is more to human nature than biology.

---

1 World Transhumanist Association, *Transhumanist FAQ 2.1.* (2003), www.transhumanism. org/resources/FAQv21.pdf (accessed 27 September 2005). We might in passing note that future transhumanist societies cannot allow completely free choice of enhancements if they wish to continue to exist. At least some individuals have to be in a state so that they can maintain the fabric of the future transhumanist society. If everyone chooses enhancements rendering them unfit for such a task, e.g. artistic instead of technological enhancement, a transhumanist society would either have to accept disintegration, or to force some, possibly future members to have specific, socially necessary enhancements.

2 D. Hume, *A Treatise of Human Nature* (London: Penguin Classics, 1985).

3 C. Takala, 'The (Im)Morality of (Un)Naturalness', *Cambridge Quarterly of Healthcare Ethics* 13 (2004), pp. 15–19.

In this paper I will argue that appeals to human nature are inevitable in many bioethical arguments and that the conception of human nature that necessarily occurs in the premises cannot be confined to a narrow conception of biological nature.

Throughout the paper I will use arguments from the enhancement[4] and transhumanism debates as examples of arguments implicitly and illicitly relying on a conception of human nature that their proponents explicitly deny. Given the size of the literature I will not provide an exhaustive survey of all possible arguments, but choose arguments for analysis that are ideal types in the sense that they state a specific position very clearly.

My arguments will not show that the fact that some intervention interferes with or changes human nature is a decisive argument against it, but they will show that using too narrow a concept of human nature invalidates the soundness of many arguments in the enhancement debate.

2. The Agreed Use of Nature

Although it might look as if some arguments in bioethical debate reject any normative import of human nature, there is one type of argument that most in the debate would agree is allowable, and that is the type of argument focusing very narrowly on the biological nature of human beings. It is difficult to deny that what is good for a human is not the same as what is good for a dog.[5] Humans can, for instance, quickly metabolize the theobromine in cocoa, and can therefore eat chocolate, whereas chocolate is poisonous to dogs because they are slow metabolizers of a range of xanthines, including theobromine. What is good and bad for you does to some extent depend on your biological characteristics and you cannot by individual fiat change this. A person might claim that poison is good for her, but given human metabolism, poison is lethal if taken in sufficient quantity.

But the pro-enhancement philosophers will immediately point out that human lives are much less determined by their biology than the lives of most other animals; and that humans have a freedom to choose their values in a way that dogs do not. As a human I am the primary judge of whether something is good for me or not, and even poison might be good for me if I want to die. The argument

---

4   I use the term enhancement instead of intervention or change, partly because it is commonly used in the literature, partly because the changes that are discussed in the enhancement debate are changes that are desired by people for themselves or for their children or more remote descendants. I may not see them as an enhancement (e.g. I don't think having your navel pierced in any way enhances you), but the person having a navel piercing must in some way desire it and believe it to be valuable for him or her. Many changes are desired primarily because they are believed to constitute enhancements.

5   This is clearly not enough to justify any claim of a radical split between human nature and the nature of other animals, e.g. primates. My project in this paper is, however, not to argue for such a radical split, but to argue that when we discuss the enhancement of humans, and possibly closely related transhumans, we do necessarily rely on a normative account of human nature. I am quite happy to allow future chimpanzee philosophers to discuss the import of chimpanzee nature for the evaluation of enhancements of chimpanzees or dolphin philosophers to ponder the normative import of being a dolphin.

is thus that if human nature has normative import, it is only human biological nature very narrowly circumscribed and such import is irrelevant for most ethical considerations. The agreement on the use of nature is thus not very far reaching.

3. *Welfare and Human Nature*

In arguments from pro-enhancement philosophers it is possible to distinguish at least four separate strands of arguments:

1. Liberal or libertarian arguments justifying why individuals have a right to pursue enhancement (and society no right to intervene).
2. Consequentialist arguments justifying why enhancement technologies should be promoted.
3. 'Defensive' arguments aimed at defusing counter-intuitive consequences of the positive arguments for enhancement.
4. Arguments based on the denial of a human nature, which oppose views sceptical to enhancement.

Implicit reference to human nature rarely occurs in the liberal arguments, apart from rhetoric concerning human inventiveness and spirit of enquiry, but it is often necessary for the soundness of the consequentialist and defensive arguments. In this paper I will concentrate on the consequentialist arguments. I will criticize the defensive consequentialist arguments first, before moving on to an analysis of the positive consequentialist arguments. In the final part of the paper I will then move on to consider whether it is at all possible to get a 'nature-less' consequentialist argument off the ground.

Let us consider the defensive consequentialist arguments first. For a liberal there is no problem in allowing competent adults to choose whatever enhancements they want for themselves, even if those 'enhancements' are actually harmful according to most conceptions of harm. The Swedish transhumanist Anders Sandberg has, for instance, proposed a principle of 'morphological freedom' protecting any choice of bodily modification.[6] In the case of enhancement of children it does, however, seem strongly counter-intuitive that the parent(s) should be allowed to freely choose 'enhancements' that are likely to negatively affect the welfare of the child. There are two defusing arguments that are commonly used to avoid this consequence; one relies on the so-called 'non-identity problem' identified by Derek Parfit and will not be discussed further here,[7] the other accepts that there is a problem and solves it by adding a rider to the liberal argument stating that parent(s) should not be allowed to choose enhancements likely to negatively affect

---

6 A. Sandberg, *Morphological Freedom – Why We not just Want it, but **Need** it* (2001), www.nada.kth.se/~asa/Texts/MorphologicalFreedom.htm (accessed 27 September 2005).

7 D. Parfit, *Reasons and Persons* (Oxford: Oxford University Press, 1984). The argument based on the non-identity problem will only be relevant if the enhancing intervention is performed before the individual in question becomes a person and if it changes what person the individual will become.

the welfare of the child.[8] An example of this kind of argumentative move can be found in a paper by the prominent transhumanist Nick Bostrom:

> One way of going forward with genetic engineering is to permit everything, leaving all choices to parents. While this attitude may be consistent with transhumanism, it is not the best transhumanist approach.
>
> [...]
>
> We currently permit governments to have a role in reproduction and childrearing and we may reason by extension that there would likewise be a role in regulating the application of genetic reproductive technology.
>
> [...]
>
> These measures have analogues that apply to genetic enhancement technologies. For example, we ought to outlaw genetic modifications that are intended to damage the child or limit its opportunities in life, or that are judged to be too risky.
>
> [...]
>
> There are grounds for thinking that the libertarian approach is less appropriate in the realm of reproduction than it is in other areas. In reproduction, the most important interests at stake are those of the child-to-be, who cannot give his or her advance consent or freely enter into any form of contract. As it is, we currently approve of many measures that limit parental freedoms.[9]

But how are we to understand this argument? What is the baseline from which we are to judge the detriment in welfare caused by 'genetic modifications that are intended to damage the child or limit its opportunities in life, or that are judged to be too risky'? We seem to be invited to perform a counterfactual comparison between the welfare of the unenhanced child and the 'enhanced' child. In principle this involves the consequentialist exercise of measuring the welfare in each time slice of the child's life and aggregating these measurements to a total welfare score for each of the two possible lives.[10] But this is impossible to do in real life. We don't have any ways of measuring the welfare in slices of actual lives, and much less any ways of measuring the welfare of slices of counterfactual lives.[11] This will not matter in cases where the 'enhancement' in question is so obviously bad that the welfare of one of the counterfactual lives is higher than the welfare of the

8   I will discuss this argument and the consequentialist arguments in the next section in terms of welfare and not preference satisfaction. Even though many pro-enhancement philosophers adhere to some form of preference consequentialism their arguments are more implausible when stated in preference satisfaction terms than in welfare terms. It is, for instance, not obvious why we should think that an individual who was more intelligent had more opportunities for relevant preference satisfaction, but more plausible that such an individual could have increased welfare.

9   N. Bostrom, 'Human Genetic Enhancements: A Transhumanist Perspective', *The Journal of Value Inquiry* 37.4 (2003), pp. 493–506 (499–500).

10   I am here bracketing the problems concerning welfare aggregation and the measurement of inequality across life times that are discussed by Temkin and Lippert-Rasmussen, and assuming that there is a non-contentious method for such comparisons, but see L. Temkin, *Inequality* (Oxford: Oxford University Press, 1993); K. Lippert-Rasmussen, 'Measuring the Disvalue of Inequality Through Time', *Theoria* 69 (2003), pp. 32–45.

11   If the enhancement causes big changes it might become problematic whether welfare measurement instruments developed to measure the welfare of unenhanced humans would be appropriate for enhanced individuals.

other in all time slices, and the lives are of equal length. In such a situation one life is clearly dominant in welfare terms and we don't need precise measurement. The pro-enhancement philosophers could claim that their defensive argument only applies to these very clear cases, but in that case it loses much of its plausibility and attraction. According to Sandberg's principle of morphological freedom I have a right to have a whole body tattoo, if I so wish. If I gave my child a whole body tattoo it is unclear whether that would make every single slice of its life worse in welfare terms.[12] It is, however, still plausible that giving my child a whole body tattoo is not something I should be allowed to do as a matter of right. The analogy to current state intervention would also break down to some extent if the argument is circumscribed to cover only clear cases, since states do intervene in various ways in much less severe cases of, for instance, 'educational deprivation'.

However, in all other cases what we do when we counterfactually compare the two lives is not some quasi-consequentialist calculation, but an overall, intuitive comparison. Think for instance of how you would judge whether ten years of life expectancy was worth giving up for a significantly increased ability to learn foreign languages. However you come to a judgement it is not by calculation. What does the intuitive comparison entail?

One way of doing it is by putting ourselves in the position of each of the two people and ask what life we would prefer to have, but it is difficult to see how we can do that without relying on our knowledge of some aspects of our own lives. But this seems to be a very subjective basis on which to judge such an important issue, since each of us only has a direct experience of a very, very limited segment of the total range of possible human life experiences. I would, for instance, be unable to judge whether having perfect pitch is a blessing or a curse. So we should probably instead try to rely on more objective, general knowledge of human life and the welfare implications of various common states of, and changes in, the human condition.

But that sounds suspiciously like relying on a quite extensive account of human nature, as well as on social context.

That social context is important in addition to human nature in determining the welfare implications of an enhancement is clearly demonstrated by the debates concerning the so-called 'social concept of disability'.[13] Although we may resist the idea that disabilities are socially constructed and that all the negative impact of disabilities can be removed by changes in society, there is no doubt that some of the negative impact of, for instance, deafness is due to social context. Likewise the putative welfare implications of enhancements will also to some degree depend on social context. Does the importance of social context undermine the line of argument pursued in this paper? It might initially seem to do so, but only on the assumption that there is a strict nature vs. social context dichotomy. But this is a clear example of a false dichotomy. We have strong reasons to believe that both

---

12 E.g. it would presumably not greatly affect the welfare of the child in the sleeping slices of its life.

13 T. Koch, 'Disability and Difference: Balancing Social and Physical Constructions', *Journal of Medical Ethics* 27 (2001), pp. 370–76.

nature and social context matter in deciding the welfare implications of many human conditions.

4. *Enhancement Without Nature*

The positive consequentialist arguments justifying why enhancement technologies should be promoted not surprisingly focus on the good consequences that enhancement will bring to those enhanced, and how widespread use of enhancement will over time lead to a better world.[14]

A typical example is the argument found in the paper by Nick Bostrom already quoted above:

> In the case of germ-line enhancements, the potential gains are enormous. Only rarely, however, are the potential gains discussed, perhaps because they are too obvious to be of much theoretical interest. By contrast, uncovering subtle and non-trivial ways in which manipulating our genome could undermine deep values is philosophically a lot more challenging. But if we think about it, we recognize that the promise of genetic enhancements is anything but insignificant. Being free from severe genetic diseases would be good, as would having a mind that can learn more quickly, or having a more robust immune system. Healthier, wittier, happier people may be able to reach new levels culturally. To achieve a significant enhancement of human capacities would be to embark on the transhuman journey of exploration of some of the modes of being that are not accessible to us as we are currently constituted, possibly to discover and to instantiate important new values.[15]

Pro-enhancement philosophers need the consequentialist arguments for two reasons. They need them to get from the position that people should be allowed to use enhancement technologies, to the position that the state should facilitate the use of such technologies and should fund their development and/or use. And they need them to try to shift the burden of proof from those who want to promote enhancements to those who want to ban them.

If enhancement technologies do not produce some kind of publicly important good (e.g. if enhancement only benefits the person being enhanced) it is difficult to argue that public authorities should spend economic or other resources on furthering them. But how are we to decide whether the world after enhancement is a better world?

I agree with Bostrom that '[b]eing free from severe genetic diseases would be good, as would having a mind that can learn more quickly, or having a more robust immune system'; but how do we get from the more robust immune system to the goodness? Well, the obvious answer is that it is good for beings like us, beings with a human nature, to get a more robust immune system because it (presumably) protects us from certain diseases. We can only see the goodness of a robust immune system in the context of a biological organism of a certain kind.

---

14 Some of the things we are being promised are very fanciful and may never come to pass, and some of the claims may therefore be claimed to be bordering on the deceptive, but a complete analysis of this aspect of the enhancement debate is outside the scope of this paper.

15 Bostrom, 'Human Genetic Enhancements', 498–99.

We therefore have to assume a specific human nature to get to the goodness of the enhancement for the individual,[16] and to the further claim that the world with the enhancement is better than the world without it. Enhancements are not free floating; they necessarily attach to individuals.

Is this implicit appeal to human nature innocuous, just an appeal to bare biology, or does it rely on a normative conception of human nature? Let us consider the parallel argument that having a more robust immune system would be good for earthworms because it would protect them from diseases and presumably enable them to live longer.[17] We would probably agree that it is good for the individual worm, but we might not think that the world was appreciably better because earthworms live longer and more disease-free lives. Why? Because earthworms do not experience their diseases or their lives in any interesting sense, and do not have life histories in the same way as humans have. The different assessments of the two parallel situations are caused by the fact that more than bare human biology has been assumed. It has been assumed that human nature involves consciousness, conscious experiences and a life history that will be improved by being protected from disease. It is only if we include all of this that we can see the extent to which a robust immune system is good for humans. But relying on the fact that humans have human life stories of a typical human type is clearly relying on a normative conception of human life and not on biology even if we take human biology to include the fact that humans have consciousness.

Here it is important to note further that this type of consequentialist argument is almost always hypothetical when it is put forward in the sense that it deals with hypothetical people having hypothetical interventions with hypothetical welfare implications in the hypothetical future. This means that the argument cannot rely on the current preferences of actual people or their own assessment of their welfare, as we can when we use consequentialist arguments in everyday life to decide how to act. Such preferences and welfare assessments could conceivably be human nature independent,[18] but the hypothetical preferences and welfare assessments are necessarily extrapolated from a conception of human nature.

5. *The Welfare of Nature-less Beings*

Let us, for the sake of argument, assume that none of the arguments above is any good and let us try to perform a nature-less consequentialist calculation. This is often necessary when we consider some of the proposed transhumanistic scenarios because they take place far into the future and contain actors that are no longer human. In one of his other papers Nick Bostrom discusses a future scenario in which a very technologically advanced successor culture has colonized the

16 We might again note that certain normative claims do seem to follow fairly straightforwardly from claims concerning human nature. I cannot make it bad for me, all things considered, to have a robust immune system simply by stating that it is bad for me, or by wishing that I had a fragile one.

17 I don't know whether earthworms actually have an immune system but that is irrelevant for the argument.

18 Although they would *ex post* inform our concept of human nature.

whole supercluster in which our solar system is situated. He invites us to consider how many human simulations could be sustained by such a culture:[19]

> As a rough approximation, let us say the Virgo Supercluster contains 10^{13} stars. One estimate of the computing power extractable from a star and with an associated planet-sized computational structure, using advanced molecular nanotechnology, is 10^{42} operations per second. A typical estimate of the human brain's processing power is roughly 10^{17} operations per second or less. Not much more seems to be needed to simulate the relevant parts of the environment in sufficient detail to enable the simulated minds to have experiences indistinguishable from typical current human experiences. Given these estimates, it follows that the potential for approximately 10^{38} human lives is lost every century that colonization of our local supercluster is delayed; or equivalently, about 10^{31} potential human lives per second.[20]

Does the loss of these '10^{31} potential human lives per second' matter? The scenario is so strange that it is difficult to marshal any definite intuitions, but the initial answer is probably 'yes'. It definitely sounds like a very bad thing to lose 10^{31} potential human lives per second. But maybe this initial judgement is too hasty and too reliant on the fact that each of us has direct phenomenological insight into what it is to lead a human life. Let us therefore, purely for the sake of argument, assume that the same computational structures can sustain any kind of processing and that there is a simple linear relationship between computational operations per second and the cognitive ability of the simulated entities.[21] If that is the case the loss could instead be 10^{30} potential lives per second of entities each having ten times the cognitive ability of an average human. Would that be as bad, worse or better? If nature-less reasoning is easy we should get a straightforward answer. My intuitions do, however, give up here,[22] but maybe this is just because the scenario is too outlandish, and not because we are doing nature-less reasoning.

Let us therefore consider a much simpler example. On your desk is a computer that because of your clever programming is the first computer in the world to reach the level of consciousness. There is an upgrade available on the market for this

---

19   I will here resist the strong temptation to comment on the quick elision of 'simulated minds' to 'human lives' in the quote, but just note that it supports the point concerning the intellectualist or mentalist bias of pro-enhancement arguments made in subsequent sections.

20   N. Bostrom, 'Astronomical Waste: The Opportunity Cost of Delayed Technological Development', *Utilitas* 15 (2003), pp. 308–14.

21   This latter assumption is probably incorrect, but the precise relationship between computational speed and the cognitive ability of the simulations is irrelevant to the argument as long as higher computational speed produces more cognitive ability.

22   However, unless the relationship between welfare and cognitive ability is also linear (or in general of exactly the same form as the relationship between computational speed and cognitive ability) consequentialist calculations must show one of these scenarios to be morally preferable, and are likely to show it to be extremely preferable given the large numbers involved even if the individual welfare differences are small. It is thus strange that our intuitions are so impotent. This example furthermore reinforces a point made above in note 8 concerning the attractiveness of the welfarist conception, because it would be even more difficult to get a handle on this comparison in preference consequentialist terms.

type of computer that will increase its processing power to twice the current level. Do you have a weighty moral reason to buy this upgrade for the computer (given that it has just become conscious it probably does not have any money to buy it for itself)?

The only way to answer this question would be to try to get inside the 'mind' of the computer and ask what it could do with the increased processing power.[23] We could try to ask it, but it might have no answer to give us, and given that it is the first conscious computer ever we will have no experience with other similar computers to draw upon. We could, of course, just assume that the upgrade would do it some good and that that would be a sufficient reason to buy it, but that would be a mistake. A belief that the upgrade would do our newly conscious computer some good is a necessary reason for buying the upgrade, but it is not sufficient because there are always alternative ways of using the money and therefore opportunity costs for every choice we make. Even if we had decided in advance to use this sum of money on the computer we would still need to know what would be best for it and therefore the best use of the money. Would such a computer, for instance, be interested in its own embodiment, as humans usually are to some degree and would a more beautiful case therefore be a more appropriate use of the money? Would it be interested in having companions of its own kind, and should we save to build it a soul mate?

The only rational, but not very satisfactory, way to answer these questions would be to try to extrapolate from the only cases we know about, i.e. human beings and to a lesser extent some animals, thereby utilizing our knowledge about human and animal natures and the connections between those and what is good for the members of the species. Later, when there are many such machines, we would be in a better position to know what would be good for them, simply because we would know more about how their nature differed from ours.

6. *Human Nature Rediscovered in Pro-enhancement Arguments*

What, then, is the implicit account of human nature that can be found in pro-enhancement arguments? As we have seen above, the human nature that is posited very often focuses on one specific attribute of human beings, the attribute that they possess intelligence. The conception of intelligence is often a very simple one, assuming that intelligence is a general all-purpose capacity for cognition and problem-solving.[24]

There are probably several reasons for this focus on intelligence. First, it is one of the mental attributes that it is easiest to imagine computers possessing.

---

23   Given the intractability of the problem we might just capitulate and assume that the computer would probably benefit from extra processing power, but this would not be very satisfactory. Maybe the increased processing power would make it more aware of the fact that it is immobile and that both input, output and power supply is not under its own control and lead it into a deep depression.

24   This is in itself a rather implausible account of intelligence. It is unclear whether there is in fact such a general capacity cutting across all or most cognitive domains, and whether an increase in intelligence necessarily leads to an improvement in problem-solving ability.

Many people will have problems in imagining that computers (or other machines) could really appreciate music or have deep emotions, but we know that they are much better at calculating than we are. Intelligence therefore suits transhumanist enhancement arguments because it can be claimed not to be human specific.

Second, if intelligence is understood as an all-purpose capacity it is easier to make a link between enhanced intelligence and enhanced welfare, and to claim that the risk of negative welfare effects from this kind of enhancement is small.

Third, many transhumanist scenarios are set in the far future and involve highly developed technologies. Increased intelligence can be seen as a stepping stone to their more rapid development.

Fourth, it provides for useful analogies with different kinds of valued social interventions aimed at improving intellectual performance, like education.[25]

Fifth, it may simply be a case of intellectualistic bias. Philosophers and other academics are probably likely to overvalue the importance of intelligence in comparison to other human attributes, and therefore likely to focus on intelligence in their arguments as one of the most important things to enhance.

There are arguments in the literature aiming to show that such a focus on intelligence is warranted. James Hudson argues on mainly consequentialist grounds that increased intelligence should be the primary aim of genetic designers and that the secondary aims should be to decrease the influence of non-rational drives and increase the capacity for experiencing pleasure.[26] As he himself recognizes, this is essentially a prescription for producing highly intelligent 'utility monsters'.

A central plank in this argument is the claim from John Stuart Mill that there are higher and lower pleasures, that intelligence is what gives us access to the higher pleasures, and that the higher pleasures are preferable. It is well known that Mill's own argument for the preferability of the higher pleasures is fallacious.[27] The argument for intelligence enhancement therefore has to rely on a reformulation of the Millian claim, and there are at least two possible reformulations in the literature. The first works best within a framework of classical hedonist utilitarianism, but can also be given a welfarist interpretation. According to this reformulation, increased intelligence gives us access to whole new ranges of pleasures, experiences, or ways of life that are simply much, much better. Just as we have access to more experiences than an elephant, so the radically enhanced will have access to more experiences than we have, and this will give them the opportunity to lead more interesting lives than ours. The remaining problem is, then, how to get from 'more interesting' to 'better in consequentialist terms' since the fact that these new lives are more interesting is mainly a function of the fact that they contain more options, not that these options can be assumed to be better in any interesting sense. There is clearly a great risk of an unjustified elision from this thin sense of 'more interesting' to 'preferable'.

25 S. Goering, 'Gene Therapies and the Pursuit of a Better Human', *Cambridge Quarterly of Health Care Ethics* 9 (2000), pp. 330–41.

26 J. Hudson, 'What Kinds of People Should We Create?', *Journal of Applied Philosophy* 17 (2000), pp. 131–43.

27 It might also be empirically false. It is, for instance, not clear whether genetic engineering aimed at enhancement of pleasures we already have would not be preferable in hedonist terms.

The second reformulation claims no direct link between intelligence and the higher pleasures, but claims that intelligence is the means by which we can create the proper conditions for the experience of these pleasures (e.g. by removing obstacles like disease or ageing). This seems relatively unexceptional although it does rely on an optimistic account of the use to which increased intelligence will be put.

A further problem with the 'intelligist' line of reasoning is that human beings are not intelligent computers who happen to have bodies and who live alone in splendid isolation. We are our bodies and although some humans can live in isolation, and some even seek it, most require social interaction to flourish. Modern neuroscience has vindicated the phenomenological analysis of Merleau-Ponty concerning the inextricable connection between our particular embodiment and our mental functions.[28] Studies by researchers like Damasio clearly show that our abilities to reason and make decisions are not purely cognitive functions but are dependent on an appropriate connection between reason and emotion and on our embodiment.[29] The focus on intelligence enhancement therefore implicates the pro-enhancement philosophers in what Damasio calls 'Descartes' error':

> This is Descartes' error: the abyssal separation between body and mind, between the sizable, dimensioned, mechanically operated, infinitely divisible body stuff, on the one hand, and the unsizable, undimensioned, un-pushpullable, nondivisible mind stuff; the suggestion that reasoning, and moral judgment, and the suffering that comes from physical pain or emotional upheaval might exist separately from the body. Specifically: the separation of the most refined operations of the mind from the structure and operation of a biological organism.[30]

It is thus questionable whether significant, isolated intelligence enhancement is actually a possibility for human beings, and if it is a possibility, whether it will actually lead to better decision making.

7. Conclusion

In this paper I have shown that many pro-enhancement and transhumanist arguments rely on a substantive account of human nature, despite the avowals of their originators. I have further shown that the reliance on human nature in the arguments is unavoidable. This indicates that human nature is here to stay in bioethical argument, at least for the time being. If pro-enhancement philosophers get their way and we move into a transhumanist society with widespread enhancement the implicit use of a conception of human nature in ethical argument might cease, but only to be replaced by the implicit or (hopefully) explicit use of a conception of transhuman nature(s).

28   M. Merleau-Ponty, *Phenomenology of Perception* (London: Routledge & Kegan Paul, 1962; French original, 1945).

29   A.R. Damasio, *Descartes' Error: Emotion, Reason and the Human Brain* (London: Macmillan General Books, 1996).

30   Damasio, *Descartes' Error*, pp. 249–50.

Chapter 4

IN WHOSE IMAGE? REPRESENTATIONS OF TECHNOLOGY
AND THE 'ENDS' OF HUMANITY[1]

Elaine Graham

1. *Introduction*

Thinking about the impact of new technologies on what it means to be human brings sharply into focus the question of relationship between technological activity and 'human becoming'. What's at stake in seeking to evaluate the significance of human engagement with its tools and technologies? Is technology a neutral instrument, an epiphenomenon to human nature? Or an integral part not only of the formation of human material culture, but of our very ontology? In this paper, I want to consider some of the debate surrounding the advent of the so-called 'posthuman condition'[2] and in particular, how the theme of *human dignity*, its endangerment or rejuvenation, features in relation to the impact of digital, biomedical and cybernetic technologies.

I will argue that technologies and the work of material fabrication are indeed a substantive, and not simply an incidental, part of being human. This is most evident in the ways in which, in contemporary debates, advanced technologies are both heralded as the means by which humanity might attain its most fundamental aspirations – the '*end*' of humanity – and perceived as constituting an urgent threat to human dignity: the end of '*humanity*'.

The challenge of the posthuman condition for theological anthropology has attracted the attention of a number of contemporary commentators.[3] Regardless of whether many of the more expansive technological developments ever materialize, the anticipated emergence of '*techno sapiens*' (Jackelén 2002) is already shaping the Western cultural and scientific imagination, not to mention the research and policy priorities of corporate and academic interests.

It would therefore seem important to trace some of the underlying values

---

1   An extended version of this paper also appears in *Ecotheology* 11.2 (2006), pp. 159–82.

2   J. Halberstam and I. Livingston (eds.), *Posthuman Bodies* (Indianapolis: Indiana University Press, 1995), p. vii.

3   Eduardo Cruz, 'The Nature of Being Human', in T. Peters and G. Bennett (eds.), *Bridging Science and Religion* (London: SCM Press, 2002), pp. 173–84; Antje Jackelén, 'The Image of God as *Techno Sapiens*', *Zygon* 37:2 (2002), pp. 289–302; Stephen B. Murray, 'Reimagining Humanity: The Transforming Influence of Augmenting Technologies Upon Doctrines of Humanity', in M. Breen, E. Conway and B. McMillan (eds.), *Technology and Transcendence* (Dublin: The Columba Press, 2003), pp. 195–216.

informing such visions of the technological future, what kinds of exemplary and normative understandings of the 'posthuman' are circulating in Western culture, and what happens when these are subjected to critique. In particular, I shall argue that aspects of theological anthropology – particularly traditions which speak of humanity as *imago Dei*, as made in the image and likeness of God – point us towards a framework of values, priorities and criteria by which the proper *ends of humanity* might be adjudicated. Yet in some respects, this paper can only set an agenda for further work in this direction, since our concepts of what it means to be human – like our very humanity itself – may not be fixed or absolute, but themselves in process, defined iteratively in relation to particular technological and cultural contexts.

2. *Endangerment or Promise?*

The impact of new technologies has had a profound impact on Western culture; and the terminology of the posthuman attempts to capture the extent to which material innovations call forth the possibility of deeper, metaphysical transformations. Yet in three particular dimensions of the posthuman future, we can discern a recurrent theme in reactions and predictions. Broadly, it is possible to trace strong threads of both 'technophobia' and 'technophilia': technologies as bearers of human diminishment, or as heralds of a brave new world of unlimited human capabilities. Three issues in particular demonstrate how particular aspects of human experience correlate with anxieties and expectations about the very future of 'human nature': embodiment, autonomy and subjectivity.

a. *Post-biological Embodiment*
If much of the debate about the impact of technologies concerns the power of new technologies either to facilitate a better quality of life or unleash Promethean powers which threaten our very humanity, then much of that attention and anxiety has come to be focused around the integrity of the body. Opinions polarize around whether it is appropriate to dream of transcending the frailties of the flesh to embrace new physical forms, or whether the end of our embodiment would fatally compromise our essential humanity.

For some commentators, the advent of new technologies will protect us against physical disease and vulnerability. Technologies, long having been the means of enabling humanity to compensate for physical limitations by providing instruments of comfort and utility, now offer the opportunities to overcome the limitations of the flesh entirely. This gives rise to the possibilities for achieving a 'post-biological' existence, as in the incorporation of artificial organs or prosthetic limbs or our transformation into 'post-bodied' individuals as we go online or experience virtual reality.[4]

Similarly, the novelty of computer-mediated communications was regarded as resting in its transformation of taken-for-granted categories of space, place and

4   See, for example, Michael Benedikt (ed.), *Cyberspace: First Steps* (Cambridge, MA: MIT Press, 1992).

time, the boundaries of the body and the interface of machine and human. In the case of the Internet, it collapses physical distance, transcends national boundaries, reinvents conventions of text and reading, transforms human interactions into the flow of information, creates new artificial environments and machinic intelligences. For some, this offers unprecedented opportunities of freedom of information and communication, the creation of a global forum of exchange and interchange.[5] Yet for others, the disembodied nature of technologies such as the Internet represents a fundamental in their 'grounding'. Despite the many benefits of computer-assisted communication, it is feared that the lack of face-to-face contact will lead to the dissolution of social bonds, as individuals descend into what Andrew Feenberg calls a 'privatized narcissism':

> The collapse of public life and the ideal of impersonal reason sets individuality loose from its institutional moorings and identity with others no longer concretized through real bonds and obligations leaves the person as a discontented spectator on his or her own life, engaged in a narcissistic and aggressive control directed towards the self and others alike.[6]

b. *Trajectory of Technologies*

One recurrent motif of modern science fiction has been that of 'creation out of control', or the fear that machines will rebel against their human masters, with catastrophic results. This anxiety is reflected in a strong and influential strand in the philosophy of technology which argues that technology has assumed a deterministic character, insofar as the imperatives of efficiency, rationalization and mass production have engulfed the priorities of human welfare. This gives rise to a sense that technological imperatives are autonomous of human design, and have an independent status in determining economic, personal, cultural and moral priorities. Technology has moved from being an instrument or tool in the hands of human agents, or even a means to transform the natural environment, and has become a series of processes or interventions capable of reshaping human ontology. It also raises the question of whether as cybernetic systems such as artificial intelligence become more sophisticated, technologies might acquire the facility to become self-programming.

Jacques Ellul speaks of the degradation of creative production to the routinization of mere 'technique';[7] Martin Heidegger argues that technology has a particular way of 'revealing' the world to its users which obscures all other possibilities than that of 'standing-reserve' – essentially, an objectification of nature which reduces everything to a commodity.[8]

For Albert Borgmann, similarly, technologies have colonized everyday life to such an extent that they have become virtually invisible. The conveniences of

5   Howard Rheingold, *Virtual Reality* (New York: Simon and Schuster, 1991).

6   Andrew Feenberg, *Critical Theory of Technology* (Oxford: Oxford University Press, 1991), p. 98.

7   Jacques Ellul, *The Technological Society*, trans. J. Wilkinson, intro. R. Merton (London: Jonathan Cape, 1965).

8   Martin Heidegger, 'The Question of Technology' [1954], in D.F. Krell (ed.), *Basic Writings* (London: Routledge, 1993), pp. 307–42.

processed food, television, text messaging and labour-saving gadgets have displaced more traditional skills and pastimes which may have involved more effort but which also constituted the essential fabric of daily life. By removing technologies from their human creators and from any sense of connectedness to their social or cultural contexts of production, Borgmann argues that we have emptied their accompanying activities of meaning. Technologies have ceased to become benevolent aids to human comfort and now serve as the taken-for-granted matrix through which all experience is mediated:

> ...the peril of technology lies not in this or that of its manifestations but *in the pervasiveness and consistency of its pattern.* There are always occasions where a Big Mac, an exercycle, or a television program are unobjectionable and truly helpful answers to human needs. This makes a case-by-case appraisal of technology so inconclusive. It is when we attempt to take the measure of technological life in its normal totality that we are distressed by its shallowness.[9]

As well as developing Heidegger's perspective on technologies' ability to reveal or conceal aspects of the world, Borgmann's analysis also carries traces of Feuerbach's and Marx's concept of 'alienation'. Technology, as the product of human labour, has itself become reified and now assumes a deterministic quality in which humans regard themselves as the objects of technological imperatives, rather than its subjects. The everyday familiarity of technologies blunts our sense of wonder at their complexity and dulls our curiosity toward their origins in the actual processes and relations of production, and their ubiquity serves as an ideological opiate to dull our moral and political sensibilities. The solution for Borgmann is to restore a sense of the origins of technologies in the human practices of fabrication and sociability, thereby returning them to the status of 'focal things'.[10]

Borgmann thus sees ways of affirming technologies in relation to human existence, despite his foreboding. However, other accounts of the future are altogether more unequivocal, regarding technologies as the benign tools of human self-actualization. Rather than envisaging the effacement of human agency, or endangering our capacity to participate in authentic existence, these more technocratic analyses are single-mindedly positive about the ability of technologies to facilitate problem-solving and advance human interests. Humans will continue to be the masters of their technological inventions, and this will enable them to dominate, subdue and eventually transcend the forces of non-human nature as well.

An example of such a perspective would be the philosophy of technology known as transhumanism. Transhumanists advocate the use of advanced medical, prosthetic and genetic techniques to extend the human lifespan, enhance human intellectual powers and improve physical and psychological capabilities. The World Transhumanist Association describes transhumanism as 'an interdisciplinary approach to understanding and evaluating the possibilities for overcoming

---

9   Albert Borgmann, *Technology and the Character of Everyday Life* (Chicago: University of Chicago Press, 1984), p. 224, original italics.

10   Albert Borgmann, *Power Failure: Christianity in the Culture of Technology* (Grand Rapids, MI: Brazos, 2003), p. 27.

biological limitations through technological progress'.[11] The possibility of the distinction between humans and machines dissolving altogether is also contemplated, as in Hans Moravec's dream of uploading human intelligence into computer software and then downloading it into an artificial environment, such as virtual reality or robotic hardware.[12]

Underlying this vision is a version of Enlightenment humanism in which the rational subject, uninhibited by the external constraints of superstition, fear or conservatism, is enabled to scale new heights of intellectual and technological achievement. Technologies are regarded as quintessentially neutral, neither good nor bad, merely instruments for achieving purposes and values enshrined elsewhere. They do not threaten or render human activities obsolete, but simply free human creativity to achieve new limits: 'Buried in the murk of the human self lies an unformed golden core, and with technology...we can tap and transform this potential.'[13]

Whilst many of transhumanists' proposed technological developments are yet to be realized, it may be more appropriate to regard transhumanism, like all other posthumanist thinking, as a kind of 'thought experiment', which, like fictional representations of technologized humanity, serves to illuminate and refract deeper hopes and fears. What makes transhumanism such a vivid example of such posthuman thinking is the way in which it articulates a particular set of humanist ideals and transposes them into the technological sphere. Transhumanists deliberately harness the aspirations of Enlightenment humanism and individualism as a philosophical underpinning for their endeavours. In its endorsement of human self-actualization unconstrained by fear, tradition or superstition, transhumanism exhibits a secular scepticism towards theologically grounded values, arguing that these serve to rationalize passivity and resignation in the face of human mortality and suffering.

Yet this is not a human distinctiveness grounded in embodiment or even rational mind per se so much as a set of qualities enshrined in a human 'spirit' of inventiveness and self-actualization:

> It is not our human shape or the details of our current human biology that define what is valuable about us, but rather our aspirations and ideals, our experiences and the kinds of lives we live. To a transhumanist, progress is when more people become more able to deliberately shape themselves, their lives, and the ways they relate to others, in accordance with their own deepest values.[14]

In this respect, transhumanism represents something of a paradox, in that to become posthuman – to be radically dependent on technology for future evolution and survival – actually entails the erosion of the very categories of bodily integrity, autonomy and personal subjectivity that define liberal 'human nature'. Furthermore, in its 'realized eschatology' of immortality and escape from

11   Nick Bostrom, 'The Transhumanist FAQ: A General Introduction, Version 2.1' (2003), 55pp, http://transhumanism.org/index.php/WTA/faq/ (accessed 9 April 2005), p. 4.

12   Bostrom, 'Transhumanist FAQ', 17–18.

13   Erik Davis, *Techgnosis: Myth, Magic and Mysticism in the Age of Information* (London: Serpent's Tail, 1998), p.128.

14   Bostrom, 'Transhumanist FAQ', p. 6.

biological contingency, transhumanism embraces a dualistic – one might say Cartesian – anthropology, in that the continuity of individuals and the human species is equated with the continuation of consciousness rather than any notion of embodied selfhood. Theologically, as Brent Waters has argued, Christian doctrine has envisaged that the griefs and limitations of the human condition take place not via a denial of death but by means of its embrace and defeat through cross and resurrection. In other words, the betterment or perfection of humanity does not occur by the negation of the body but by embracing and redeeming corporeal humanity in all its vulnerability.[15]

In addressing the claims of transhumanism, therefore, it is important to recognize its implicit truth-claims regarding normative and exemplary human nature. It is perhaps the starkest example of a posthuman doctrine seizing upon selected aspects of cultural understandings of what it means to be human – in this case Enlightenment liberal individualism – and projecting these as universal and inevitable aspirations of technological endeavours. Yet as Katharine Hayles argues, the dilemma is to affirm the beneficial potential of technologies without being seduced by their Gnostic or Platonic tendencies:

> If my nightmare is a culture inhabited by posthumans who regard their bodies as fashion accessories rather than the ground of being, my dream is a version of the posthuman that embraces the possibilities of information technologies without being seduced by fantasies of unlimited power and disembodied immortality, that recognizes and celebrates finitude as a condition of human being, and that understands human life is embedded in a material world of great complexity, one on which we depend for our continued survival.[16]

c. *The Death of the Subject?*

A fundamental fault-line exists at the heart of all these perspectives on technology. Does the 'posthuman condition' represent promise or endangerment, mastery or extinction? If 'critical posthumanism' is asking the question 'What does it mean to be human in an age of technologies?' where is it to look for an answer? If *homo sapiens* is to be succeeded by *techno sapiens*, what happens to our accounts of human subjectivity, and in particular the basis on which we might ascribe it a moral status? In the face of the alleged dissolution of any normative concept of human nature, how are criteria for differentiating between legitimate and illegitimate technological appropriations to be articulated?

These are questions regarding the future trajectory of human engagement with technologies, and whether the transition from *homo sapiens* to *techno sapiens* will propel humanity towards greater knowledge and prosperity, or diminish human uniqueness to the point of obsolescence. Yet how seriously should we take these posthuman claims that technology is reshaping our deepest experiences and understandings of what it means to be human?

---

15 Brent Waters, 'From *Imago Dei* to Technosapien? The Theological Challenge of Transhumanism', unpublished paper, 2004.

16 N. Katherine Hayles, *How We Became Posthuman: Virtual Bodies in Cybernetics, Literature and Informatics* (Chicago: University of Chicago Press, 1995), p. 5.

3. *Post/human Technologies*

Philosophical speculation about the relationship between humanity and technologies has a long history, and current thinking can be placed in a long legacy of debate.[17] Nevertheless, certain advances over the past half-century – most particularly the construction of the world's first stored-memory computer in 1948 and the identification of DNA in 1953 – do seem to represent an unprecedented intensification of technoscientific[18] potential. The shift into a so-called 'posthuman' sensibility thus signals a recognition that technology as 'other' to human existence is now assimilated into everyday human functioning: in a literal sense, in terms of the possibilities of devices such as cochlear implants, intraocular lenses, heart pacemakers and other artificial organs or prostheses becoming a permanent part of human physiology, and in a more existential sense, insofar as most of the population of the 'first world' are completely dependent on technologies such as computer-mediated communications, broadcasting media, transport infrastructure and generation of power.[19]

The posthuman has thus become a shorthand metaphor for talking about a number of interrelated issues. First, the scientific and medical prospects for *enhancement of human physical and intellectual powers* by means of cybernetic implants or genetic modification, including increased longevity, resistance to congenital or other illnesses, improved cognitive skills. Secondly, as a corollary to this, the posthuman has come to be identified for many with the emergence of a *new phase of human evolution*, albeit technologically driven (and the result of human choices) rather than driven by the forces of natural selection.

The danger is, however, that this brand of posthuman thinking simply adopts a taken-for-granted assumption that some time before the advent of biomedical, cybernetic and genetic technologies the boundaries between the human/non-human, natural/artificial, organic/technological were fixed and axiomatic, and to be human rested on a clear 'ontological hygiene' of essential qualities.[20] Yet the boundaries have always been contested, and attempts to define the 'human' in relation to the 'non-human' is a work of exclusion, a denial of our entanglement, our complicity, with the world of our tools, technologies and environments. In a digital and biotechnological age, therefore, it is impossible to speak of a pristine, unadulterated 'human nature' without considering the ways in which humans have always, as it were, co-evolved with their tools and technologies. Yet this may be no more than a reminder that as the creators of material cultures, via the pro-

17   Robert C. Scharff and Val Dusek (eds.), *Philosophy of Technology: An Anthology* (Oxford: Blackwell, 2003).

18   Use of the term 'technoscience' is frequently attributed to Bruno Latour, whose work on the social contexts of scientific research implicitly refutes any notion of the separation of science and technology into 'pure' and 'applied' forms of knowledge. The priorities of practising scientists are shaped by wider cultural factors, not least the availability of funding; and the logic of technological developments often exert defining influences over cultural institutions and behaviour, not least in the harnessing of scientific endeavours to corporate commercial interests.

19   Elaine L. Graham, *Representations of the Post/Human: Monsters, Aliens and Others in Popular Culture* (Manchester: Manchester University Press, 2002).

20   Graham, *Representations of the Post/Human*.

cesses of transforming or constructing the world, humanity cannot but refashion *itself* as integral elements of that world.

> For what does it mean to say that our context is *now* posthuman? What does it mean to say that the human is intertwined with non-humans *as never before*? When was the human *not* inextricably entwined in material, technological and informational networks? When was the human ever just 'itself'?[21]

There is a tendency to regard 'technology' as merely synonymous with advances in computing, genetic engineering and the industrial age. But the root of the word, the Greek *techne*, or 'craft' or 'skill', reminds us that technologies are both more basic and more ubiquitous within human culture. Indeed, they may be the very facility that distinguishes human beings from other non-human animals, insofar as they make possible sophisticated cultures, enable human beings to sustain long-term inhabitation of difficult environments, as well as being the very medium through which cultural achievements – such as language, visual arts, material culture, economic production, exchange and consumption, as well as religion – are embodied. Technology is thus both a body of knowledge, in terms of accumulated know-how, and the material instruments and machines through which such knowledge is practised.

All human societies may be said to be 'technological', therefore, insofar as they deploy tools, devices and procedures to assist human living. Such technologies also actually fuel human evolution, as humans adapt to changing conditions by inventing technologies that enable us to maximize our chances of survival and development in previously hostile or problematic contexts. Yet those technologies, arguably, are more than mere appendages to autonomous human reason. They actually affect our experiences and apprehensions of what it means to be human so that we cannot conceive of ourselves independent of our tools and technologies.

In my view, therefore, talk of the 'posthuman' is not simply a marker of an inevitable process of succession, nor even a clearly defined set of characteristics, so much as an invitation to consider how cybernetic, digital and genetic technologies call into question the deepest assumptions underlying our notions of normative and exemplary humanity. This strand of what Bart Simon has termed 'critical posthumanism' might best be described as indicating the *reconstitution* of the subject.[22] Like much of poststructuralism, which exposes the artifice of modernist constructions of the subject, 'post' unsettles its suffix, serving to betray its fictional and contingent status. The posthuman condition is one in which the boundaries of humanity are interrogated; not so much a singular epoch, then, or a particular evolutionary vision, but more a critical lens.

The posthuman is therefore less a 'condition' than an *interrogation*; and for this reason, I have chosen to adopt the notion of the 'post/human' – the forward-slash or oblique serving as a reminder that our delineation of human nature may be

---

21   Bruce Braun, 'Querying Posthumanisms', *Geoforum* 35 (2004), pp. 269–72 (271).
22   Bart Simon, 'Toward a Critique of Posthuman Futures', *Cultural Critique* 53 (2003), pp. 1–9.

more a question of *boundaries* than essences.[23] As we encounter anew ourselves and others, so we embark on theoretical realignments on issues such as identity, subjectivity, embodiment, reproduction, life, death, human creativity. Similarly, given the blurring of human/machine (as in the figure of the cyborg, or the multiple, role-playing, virtual self, dependent on computer communications for its realization) we can no longer claim just what exactly constitutes the human subject that might be enhanced or propelled into cyberspace; in fact we cannot ontologically separate humanity from its technological medium. 'Human nature' ceases to be a fixed category and re-emerges as a constantly changing set of possibilities and configurations.

In thinking about the future of human nature in a technological age, therefore, it may be more appropriate to acknowledge that humans are capable of building and inhabiting all kinds of worlds – some material, others virtual – but that they are all bound by the relations of their production, albeit working according to different conventions. Who and what 'we' are or become is thus intimately linked to the conditions under which we engage with technologies, and to what end. Neither the 'transcendence' of the technological sublime nor the 'determinism' of disenchantment fully captures the complexity and reflexivity of the necessarily hybrid nature of post/human experience.

4. *In Whose Image?*

I have identified some of the values about what it means to be human which are fuelling the anxieties and hopes surrounding the emergence of cybernetic, genetic and digital technologies. Despite the predominance of secularism in modern science, it is intriguing to identify a number of ways in which religious discourses still imbue many of the considerations of what should constitute our post/human future. This does not simply apply to those who regard technologies as responsible for the *disenchantment* of the world in the name of spiritual values, but also to those who deploy theological language to justify technologically driven human evolution. Yet the ways in which religion is evoked to support this project are also problematic, as critical dialogue between some of these tendencies and Christian theological anthropology will indicate.

Narratives of transcendence, disembodiment and mastery, masquerading as eternal, enduring, universal 'religious' instincts, are elevated as exemplary ideals. Hence my reference earlier to the aspirations of transhumanism as a form of 'realized eschatology', despite the movement's antipathy to religion. In assuming that the end of humanity (as in its ultimate destiny) is to become like God, or the gods, however, an apparently secular doctrine of humanity continues to draw on a traditional theological motif: that of the *imago Dei*. Humans are obliged to overcome those elements of their nature that are not divine (mortality, embodiment, contingency) in order to aspire to the true marks of divinity. Yet in fact this implicit religious vision bears little relation to many of the conceptions of the *imago Dei* that Christian theology has traditionally developed.

---

23   Graham, *Representations of the Post/Human.*

Whilst there are many strands to the Christian notion of humanity made in the image or likeness of God,[24] it is perhaps in relation to God as Creator that alternative themes in relation to human *techne* and technological endeavour might be made: values which do not deny the significance of human inventiveness, but which locate the *telos* of any such activity in a very different set of divine – and therefore human – attributes.

It is certainly important to see *techne* as spiritually and theologically worthwhile, and to affirm that that quintessential aspect of our very humanity is realized in and through our relationships with our tools and technologies. Technologies may be regarded as tools by which our capacity to create is brought to fruition, a way of resolving problems, of extending human capabilities and meeting needs. Yet technologies are not merely means to the ends of faster communication, greater crop yields or secure dwellings: they are objects of beauty as well as utility, artefacts and creations in their own right. They are thus both the vehicles and the expressions of human creativity, an integral part of our very human nature; or in the words of Philip Hefner, a necessary means of 'human becoming'.[25] In that respect, human beings may be said to be legitimately enacting the image or likeness of God when engaging in technological innovation. The possibility of human evolution continuing through technologies is not to be rejected. Rather, it is possible to argue that a quintessential aspect of our very humanity is realized in and through our relationships with our tools and technologies.

So on the one hand, it is important to see technologies not simply as mere instruments of doing and making, but as vehicles of human development and self-realization. What happens, as in the debate on the post/human, however, when human beings themselves become the objects of their own self-design, or when biological life can be modified and engineered in ways which 'transcend' or surpass the workings of unassisted nature? Some further qualifications may therefore be necessary in relation to the potential and limits of human creativity.

One approach within Christian theology would be to affirm human inventiveness whilst arguing that it is akin to divine creativity only by analogy. Humans are not capable of creating the universe *ex nihilo*, even though they may be capable of creating life by biological and technological means. This image of humanity as 'created co-creators'[26] expresses it well, not least because it encapsulates a theological anthropology of likeness and affinity to divine activity, yet limits any notion of humanity's complete or literal equivalence to God. Humanity is both creator and creature, affirming yet also conditioning claims to the autonomy of human self-design. It is suggestive of an understanding that the

---

24   Such as, for example, thinking of humanity as possessing certain qualities or capabilities (such as creativity or rationality), as essentially relational beings (after Trinitarian understandings), or as called to particular responsibilities and activities (such as stewardship), all of which are human expressions of divine nature. See Noreen Herzfeld, *In Our Image: Artificial Intelligence and the Human Spirit* (Minneapolis: Fortress, 2002).

25   Philip Hefner, *Technology and Human Becoming* (Minneapolis: Fortress, 2003), p. 43.

26   Hefner, *Technology and Human Becoming*, p. 260.

final ends of creation cannot be subsumed either to technological imperatives or contingent human ambitions.[27]

Similarly, talk of humanity as in some degree self-constituting via its own technologies, of being capable of influencing the course of its own development, is to fall prey to what we might term 'hyper-humanism': a distortion of modernity's faith in the benevolence of human reason, producing the hubristic belief that humanity alone is in control of history. According to this model of technology and human becoming, humanity comes to consider itself the supreme source of value without reference either to its common origins in nature, or of its relationship to a horizon of otherness – such as a divine Creator. The notion that humanity can seek redemption via technocratic means alone – whether this would be via genetic modification, cybernetic or prosthetic enhancement, even social engineering – represents the ultimate in the elevation of human perfection as the apotheosis of our technological, moral or political endeavours. Hyper-humanism may even be regarded, in theological perspective, as a form of *idolatry*: the elevation of human, finite creation as ultimate reality; a confusion of the fabricated with the divinely created.

A Christian theological anthropology, by contrast, sees things differently, not least in its eschewal of a symbolic of transcendence premised on omnipotence, individualism and immortality. Rather than regarding the immanent, embodied, material world as an impediment to genuine spirituality, or – reminiscent of Gnostic world-views – as the profane, flawed, pale reflection of the authentic divine world, this vision sees it as the very realm of divine–human encounter. 'Whatever it is, salvation will be an affirmation of the essential finitude of human nature, not an escape from it.'[28]

Therefore, *imago Dei* in theological anthropology is an expression of the potential of human beings to grow into likeness of God in Christ, and of Christ as the model for the perfection of humanity. Yet to aspire to *imago Dei*, to see human fulfilment in the image of God as revealed in Christ, properly leads to humility rather than self-aggrandisement. To 'become Divine' in Christian terms is to follow a pattern of divine *kenosis* in Christ, an acknowledgement of the unconditional, parental love of the Creator, and a commitment to absolute trust in the values of the Kingdom and not those of earthly powers. Again, this is not to argue for a rejection of technologies per se, but to recognize that if they are to be fully a part of human ends then they are necessarily complicit in activities which build worlds of moral value as well as those of material objects – worlds which take account of human existence in wider contexts of responsibility and mutuality with one another, with animals, environment and other living organisms. This is in part because if humans are *created* beings, then they share that status with the rest of non-human nature; but an acknowledgement of such a common affinity grounded in creatureliness is inconsistent with human engagement with

---

27   Peter Scott, *A Political Theology of Nature* (Cambridge: Cambridge University Press, 2003).

28   David Kelsey, 'Human Being', in P. Hodgson and Robert King (eds.), *Christian Theology: An Introduction to its Traditions and Tasks* (London: SPCK, 1998), p. 144.

nature, non-human animals and technologies based on principles of mastery, control and appropriation.

Similarly, it is not necessarily fatalism or passive resignation to admit the finitude of creation or human links to nature and responsibility. This is where cyborg anthropology differentiates itself from transhumanism. The recognition of the hybridity of humanity in an age of technology, and the interdependence of nature, humanity and technology, comes from a sense that we cannot evade either our biological origins or our technological ambitions. Rather, an ethical responsibility begins with that very ambivalence and complicity, whereas as I have argued, transhumanism seeks to escape all limits of scarcity, the consequences of human actions on non-human nature and questions of access and distribution.

The *imago Dei,* like transhumanism, speaks of the capacity of humanity for ultimate perfection; but whereas the former speaks of that as bounded only by the boundaries of human imagination or capability, the latter identifies human potential and growth in *imitation Christi.* The difference lies in the source of the benchmark for our ideals of human potential and human dignity. Transhumanism tends to identify perfection with physical and intellectual functioning: freedom from vulnerability, even mortality; but theological anthropology stresses wisdom, compassion and self-emptying.

To speak of humanity created by God, in God's image, thus provides an alternative horizon by which human value is located as transcending human utility or objectification and restored to a deeper irreducibility. It also propels us towards a theological anthropology which talks of human creatureliness, a dependence on our common origin in divine purpose rather than self-design.

5. *Conclusion*

If 'the posthuman condition is upon us',[29] then the significance of this goes beyond the impact of specific technologies to embrace deeper questions about the future of humanity itself. In surveying the range of responses to the post/human, we have seen how the cybernetic, biotechnological and digital age is regarded as simultaneously 'endangerment' and 'promise' to human integrity. Yet crucially, this only serves to illuminate implicit philosophies of the relationship between human agency and creativity, and its products: our tools, artefacts and technologies.

I have emphasized how different perspectives on the nature of technologies reflect alternative understandings, and in particular the contrasting visions of 'enslavement or liberation' within their visions of the post/human future.[30] I have argued, however, that neither a model which envisages humans as helpless puppets in the face of technological determinism, nor that which regards technologies as mere instruments at the disposal of the liberal humanist ego, are adequate to describe the relationship. We cannot essentialize technologies, either *reifying* them as possessing powers beyond our control, or *deifying* them as the sure

---

29 Halberstam and Livingston, *Posthuman Bodies,* p. vii.
30 David Cooper, 'Technology: Liberation or Enslavement?', in R. Fellows (ed.), *Philosophy and Technology* (Cambridge: Cambridge University Press, 1995), pp. 7–18.

means to our future salvation, as if they had been created outside social contexts, political and economic choices, or even independent of human agency. Instead, I have argued for a rather more *reflexive* understanding of technologies, insofar as they emerge from, and reflect, social, political and economic priorities in terms of design and deployment, and are therefore never culturally, morally or politically neutral.[31] We return instead to the fundamental issue at the heart of critical post/humanism: technologies are not neutral tools or instruments, but shape our very engagement with the world, and potentially our very ontology.

This serves to remind us that important value-judgements lie at the very heart of our engagement with technologies: what kinds of humans we may become; how humanity will use the resources, commodities and artefacts granted to it; questions of access, equity and distribution. These are, ultimately, questions of how we use our creative abilities – abilities to build and inhabit all kinds of worlds, material, virtual and imaginary; but worlds which, nevertheless, can either enhance or diminish the precious commodity of human dignity.

I have argued that the notion of humanity as created in the image of God provides a useful benchmark for thinking about the future of human nature in a technological world. An understanding of *human* creativity as participation in *divine* creativity affirms the goodness of our inventive abilities: and it is important to celebrate our capacity to be 'builders of worlds', tool-makers and tool-users, and see this as part of what makes us distinctive as a species. Yet to conceive of the 'transcendence' of God as that which understands God as 'radically other' does not necessarily imply a flight from the material. It may simply indicate the irreducibility of the divine to human self-interest. As such, it would serve as an important antidote to the narratives of hyper-humanism in which the 'will to power' is used as the rationalization for aggrandizing versions of corporate interest and transhumanist superiority. 'Transcendence' needs therefore to be linked with materialist and incarnational, rather than quasi-Gnostic, doctrines of creation, signalling an understanding of humanity made in the image of God which, far from being granted licence to conquer and subdue, recognizes that human *creativity* is always already framed by human *creatureliness* and interdependence. In turn, this offers a necessary reminder that our technologies are ultimately not our own, that our inventions, like the whole of creation, only make sense when offered up as part of a larger, divine purpose.[32]

Although the prospect of humans being mixed up with nature, machines and non-human animals may seem disturbing, it is, I believe, simply a reflection of the fact that human beings have always, as it were, 'co-evolved' with their environment, tools and technologies. By that I mean that to be human is already to be in a web of relationships, where our humanity can only be articulated in and through our environment, our tools, our artefacts, and the networks of human and non-human life around us.

---

31   Feenberg, *Critical Theory of Technology*.
32   See Scott, *A Political Theology of Nature*, for a discussion of how an eschatological understanding of creation also redirects discussion of human *telos* in a technological age, in that creation cannot be brought to fruition by human effort alone, but awaits Divine completion.

Yet the very hybridity of human being simply compounds the difficulties of making appeals to human dignity on ontological grounds. Human nature is not 'a template untouched by social contingencies and historical becoming';[33] indeed, theological anthropology calls for an incarnational, relational, performative model of human personhood that regards the material world and history as the proper sphere of creation and redemption, and of humans as 'created co-creators'. It also means, I think, that we cannot afford to be afraid of our complicity with technologies, or fear our hybridity, or assume that proper knowledge of and access to God can only come through a withdrawal from these activities of world-building. What it means to be human, what is happening to the material world, are not matters which divert us from the true task of spiritual reflection and Christian living, but their very preconditions.

33 Scott, *A Political Theology of Nature*, p. 261

Part II

MEDICALIZED HUMANS

Chapter 5

In the Waters of Babylon: The Moral Geography of the Embryo[1]

Michael S. Northcott

On 11 August 2004 the Human Fertilisation and Embryology Authority (HFEA) granted the first UK licence to scientists in the University of Newcastle for therapeutic human cloning. The licence allows scientists to create human embryos by inserting the nuclei from human skin or stem cells into human eggs, a procedure which they announced in May 2005 had been successful.[2] The purpose of this procedure, according to the chair of the HFEA, Suzi Leather, is as follows:

> This licence allows scientists to create human embryos by inserting the nuclei from human skin or stem cells into human eggs. In the UK, research on human embryos is only permitted for certain purposes. The purpose of this research is to increase knowledge about the development of embryos and enable this knowledge to be applied in developing treatments for serious disease. This research is preliminary, it is not aimed at specific illnesses, but is the foundation for further development in the treatment of serious disease.[3]

The goal of the research is described by the HFEA as being to 'increase knowledge about the development of embryos and enable this knowledge to be applied in developing treatments for serious diseases'.[4] This licence allowed the Newcastle Centre for Life to become the first laboratory in the UK to undertake instrumental research leading to the creation of cloned human embryos as means to the end of enhancing scientific knowledge. Since 2004 a further three licences have been granted to scientists in Newcastle, Edinburgh and Cambridge to undertake similar research. The granting of the first licence provoked extensive public debate and was successfully challenged in the High Court although the government overturned the first ruling on appeal.[5]

1   I am grateful to Nick Adams, Bernd Wannenwetsch and Sam Wells, and to the co-authors and editors of the present volume, for their insightful comments on earlier versions of this paper.
2   'UK Scientists Clone Human Embryo', BBC News, 20 May 2005, http://news.bbc.co.uk/1/hi/health/4563607.stm.
3   Human Fertilisation and Embryology Authority, *HFEA Grants the First Therapeutic Cloning Licence for Research*. Press release, 11 August 2004.
4   For an account by the Newcastle team of the scientific reasoning behind embryonic cloning see Miodrag Stojkovic, Majlinda Lako, Tom Strachan and Alison Murdoch, 'Derivation, Growth and Applications of Human Embryonic Stem Cells', *Reproduction* 128 (2004), pp. 259–67.
5   Peter Herissone-Kelly, 'The Cloning Debate in the United Kingdom: The Academy Meets the Public', *Cambridge Quarterly of Healthcare Ethics* 14 (2005), pp. 268–79.

The production of embryos in Petri dishes is, of course, not new. In Vitro Fertilization (IVF) has the same goal, using human semen as its means. The innovation in the Newcastle study is a radical change in means: the new licences allow scientists to fertilize human oocytes from cells or cell cultures taken from other humans so creating cloned embryos by parthogenesis. The principal technique involved, cell nuclear transfer, was devised by Ian Wilmut and led to the creation of the first cloned mammal, the sheep he named Dolly.[6] The reason given by the HFEA for the so-called 'responsible use' of this technology for specifically human cloning is that it will assist in advancing scientific knowledge with the hope that this will assist in the therapeutic treatment of disease, and in particular in the production of embryonic stem cells which may be used in new genetic therapies. The major moral objection to this novel therapy is that it results in the creation of an embryo which, in principle, has the potential to become a child and subjects it to experimental procedures as means to the end of advancing knowledge. Although the law in Britain forbids the implantation of cloned embryos into the uterus, which it defines as 'reproductive cloning', 'therapeutic cloning' nonetheless represents a clear moral threat to human dignity and integrity, as opponents of the HFEA's decision have argued both in court and in other forums.[7]

The advent of so-called 'therapeutic' cloning depends not only on Wilmut's success in nucleic cell transfer in the cloning of mammals but also on technical 'advances' in the pursuit of Assisted Reproductive Technologies (ART) and in particular on the extraction of human eggs from women's ovaries, their fertilization in Petri dishes, and their storage in freezers as permitted in pursuit of the treatment of infertile couples.[8] It is precisely the eggs that can now be described as 'surplus' and 'spare' (as well as those damaged by the IVF treatment process) that provide the material for cloning and embryonic experimentation.[9] Such experimentation would not be possible if eggs and embryos did not already exist outside the uterus: outside the maternal womb. The procedures associated with IVF disembed embryos from their natural and social environment in the human body and in kinship relations, and as Oliver O'Donovan was among the first to suggest, they consequently become artefacts, objects of human making and experimentation.[10] Once their natural contexts no longer determine the kinds of being that they are to be, their 'kinds' are determined through acts of human willing or choice. It is

6   For the paper describing the breakthrough see I. Wilmut, A.E. Schnieke, J. McWhir, A.J. Kind and K.H.S. Campbell, 'Viable Offspring Derived from Fetal and Adult Mammalian Cells', *Nature* 385 (27 February 1997), pp. 810–13. For ethical reflections on the cloning of Dolly see Michael Northcott, 'Concept Art, Clones and Co-Creators: The Theology of Making', *Modern Theology* 21 (2005), pp. 219–36.

7   There was a legal challenge to the HFEA licence by Josephine Quintaville, made on behalf of the campaign group Comment on Reproductive Ethics, which was upheld by the High Court but overturned on appeal: see further Herissone-Kelly, 'The Cloning Debate in the United Kingdom'.

8   Ian Wilmut has also been granted a licence by the HFEA to pursue therapeutic human cloning: Gretchen Vogel, 'Stem Cell Research: Cloning Pioneer Heads Toward Human Frontier', *Science* 298 (4 October 2002), pp. 37–39.

9   Editorial, 'Disease Insights from Stem Cells', *Nature* 422, no. 6934 (24 April 2003), p. 787.

10   Oliver O'Donovan, *Begotten or Made?* (Oxford: Oxford University Press, 1984).

true that 'natural contexts' are also cultural products, but we do not quite 'will' or 'choose' these; they are deeply rooted in long histories. By contrast, the new willing and choosing that determine 'what kind of thing' a laboratory embryo is have a quite different and much shorter history: they are a product of professional judgements by technical experts writing reports intended to influence legislation. These experts do not have parental duties or instincts, and the range of persons who make choices about the fate of embryos extends beyond the parents, who are in any case no longer the parents, but merely the 'donors', of the embryos, as the following case illustrates.

In 1998 a widow in Toulouse contested with a hospital over the status of two frozen embryos which remained from a programme of IVF treatment during the course of which the husband died in a car accident while on his way to visit his wife in the city where the treatment was taking place. His widow insisted that the two frozen embryos should be inserted into her uterus, as had other such embryos before the death of her husband, as she wished still to carry his children and as his intention before his death was that she should bear his child. The hospital refused and the case went to court. The legal grounds on which the mother's request was turned down by the court included the judge's concern at the moral and psychological effects on a child subsequently born who would in effect be an orphan, without the possibility of knowing its biological father. The judge believed this would be wrong.[11] This ruling is remarkable for two reasons. First, it is surprising, though by no means unintelligible from a Christian perspective, to hear a European court in the late twentieth century judge that the proper context for the nurture of children is a two parent, male and female household: the opinion would make all IVF procedures legally questionable which did not involve the sperm and egg of two known and cohabiting persons. Second, it reveals the role of a network of persons – doctors, lawyers, judges – who decide the chances of life of a 'spare' embryo, in addition to parents or 'donors'. That this network of persons is responsible for deciding on the potential life or death of an embryo is an indication of the changed social context of the embryo in the freezer or the Petri dish; suspended in sterile fluid in these novel terrains the embryo is in exile from its natural home in the womb of a woman and is subject to alien powers, which is to say powers not naturally and normally operative in the matter of the fate of an embryo in a human body.[12] As Novaes and Salem comment:

> Deprived of its 'natural' setting and traditional references to the woman's body, the embryo seems to be at present immersed in an ambiguous 'no (wo)man's land' with an increasing number of actors who, for different reasons, feel responsible for its destiny…frozen embryos (and even foetuses) may be found in a limbo-like space characterized by internal disputes and by incipient normative references regarding the question of how the arguments and criteria of actors who feel concerned with an embryo's future should be ranked… In this sense,

11 The case is described more fully in Simone Batement Novaes and Tania Salem, 'Embedding the Embryo', in John Harris and Søren Holm (eds.), *The Future of Human Reproduction: Ethics, Choice, and Regulation* (Oxford: Clarendon Press, 1998), pp. 100–26.

12 Novaes and Salem, 'Embedding the Embryo', p. 126.

the embryo's moral status and concrete destiny reveal, and will always reveal, arbitrary social choices.[13]

By disembedding the embryo from its natural context in the womb, IVF has paved the way for the existence of a whole new race of potential lives in Petri dishes or freezers. In the case of cloned embryos, these potential lives come to lack even one intending biological parent. Cloned embryos are even more morally and socially disembedded from the human body, from kinship relations and natural reproduction processes than those embryos used in IVF treatment and this disembedding is presumably intentional. If cloned embryos in some senses belong to particular 'parents', from whom the eggs or DNA derive, then their use by scientists 'to increase knowledge about the development of embryos' in the pursuit of therapies would be problematic.[14] It is precisely their disembedding from any particular human body or parent which facilitates their use as means to a potential therapeutic end.

Some feminists welcome the decontextualization of reproduction from sexual relationships and the womb. Dion Farquhar argues that:

> [t]he separation of reproduction from sex forces an acknowledgement of the historicity and constructedness of reproduction. Reproduction becomes historically situated. Despite her submission to invasive, stressful, expensive, and sometimes painful medical intervention, the female ART user need fake no orgasms nor provide emotional or sexual service to her impregnators in order to reproduce.[15]

On this account Assisted Reproductive Technology (ART) represents a further stage in women's liberation, enabling women to have babies independently of the putatively invasive demands of men on their bodies, emotions and lives.

Other women, including some feminists, reflect more critically on the technological invasiveness of ART and its effects on their bodies and sexual relationships. As an Australian woman in an IVF programme observes:

> My husband thinks I have lost interest in him. It is not that. It is just that that whole area is so painful I want to deny it exists. The other night while we were making love I thought 'this isn't something special between the two of us, it is something which involves all these other people'.[16]

Added to the negative effects of the procedures on the relationships and sense of embodiment of those who endure these procedures are the negative effects which arise from the considerable failure rates of IVF procedures which vary between 50 and 85 per cent. The effects of receiving treatment which is unsuccessful are often very negative:

> When you are on the program you are set up to be a parent...You are very sup-

13   Novaes and Salem, 'Embedding the Embryo', p. 126.
14   HFEA press release, 11 August 2004.
15   Dion Farquhar, *The Other Machine: Discourse and Reproductive Technologies* (New York: Routledge, 1996), p. 190.
16   Ana Murdoch, 'Off the Treadmill – leaving an IVF Programme Behind', in Jocelynne A. Scutt (ed.), *Baby Machine: Reproductive Technology and the Commercialisation of Motherhood* (London: Green Print, 1990), p. 70.

ported by the team. It is like a little world away from the real world. You are encouraged to be the person you think you want to be, that is a parent, and not what you are, which is infertile. IVF pressures the community into thinking that anyone can now get pregnant. There is an emphasis that if anyone wants to be happy they must go and have this baby transplanted into them: 'Medicine will make me happy.'[17]

Even in a purely consequentialist frame, these unhappy outcomes would seem to act as arguments against IVF: if net human happiness is not advanced by IVF – that is, if more women leave such programmes unhappy or childless than those who leave them happy or successfully pregnant – then it is hard to see how even a consequentialist might continue to argue plausibly for the morality of these procedures.

It is argued that therapeutic cloning is not like IVF. Its advocates suggest that, unlike IVF, parents are not directly involved and therefore it will not have these kinds of unhappy outcomes. The welfare of potential mothers is not at issue, as the only role they will have in therapeutic cloning is as egg donors who have (at least in the UK) been asked to consent to the experimental use of their eggs. But that the welfare of mothers is undiminished by therapeutic cloning is open to doubt. If the imaginations of laboratory staff are constrained today, the authors of yesterday and today show themselves well aware of issues of this kind. The classic account of a genetically modified society in which all persons are born by embryo pre-selection is Aldous Huxley's *Brave New World*.[18] The society he describes is biologically stratified such that embryos are pre-selected to play their parts as virtual slaves or members of a ruling intelligentsia by the design and will of the state. Such a society is clearly one in which men and women are reduced to what Giorgio Agamben calls 'bare life' in such a way as to produce a new kind of tyranny which enslaves the human body in an all-encompassing project of control.[19] In Margaret Atwood's novel *Oryx and Crake*, human civilization reaches a point of apocalyptic breakdown because of a failed experiment in the genetic modification of humans.[20] A new genetically modified virus is inadvertently released from a laboratory and the majority of the human populace, including most mothers, die as a result.[21] Those few that survive are left with their memories of what went before which include, for the novel's narrator Snowman, a memory of a mother's anger, and depression at the corporate manipulation of life for profit which had been hers and her husband's livelihood when they both worked for a pharmaceutical company. Both these novels indicate that the clinicians and others

17   Isobel Bainbridge, cited in Murdoch, 'Off the Treadmill', p. 67.
18   Aldous Huxley, *Brave New World* (London: Chatto and Windus, 1932).
19   Giorgio Agamben, *Homo Sacer: Sovereign Power and Bare Life* (Stanford, CA: Stanford University Press, 1998).
20   Margaret Atwood, *Oryx and Crake* (London: Bloomsbury, 2003).
21   Atwood's scenario is not far fetched. In the paper setting out the scientific reasoning behind the creation of embryonic stem cells, the Newcastle team headed by Alison Murdoch note that the use of 'animal-based ingredients' in the culture of human embryonic stem cells 'incurs a risk of cross-transfer of pathogens': Stojkovic *et al.*, 'Derivation, Growth and Applications of Human Embryonic Stem Cells', p. 259.

who participate in techniques of egg extraction, embryo fertilization and storage
are enacting a narrative about the nature of persons which is morally charged and
which has potentially destructive implications for children and adults in societies
which embrace it. [22] According to this narrative embryos or foetuses are apersonal
and therefore not deserving of respect as persons. Such an account of what it is
to be a person conceives of the person as a mere container of qualities – a functioning brain, powers of reason, communicative abilities – which, when lacking,
mean that what traditionally may have been judged to be constitutive of persons,
including embryonic or foetal life, is not judged as deserving of the respect due
to persons even though persons could not exist who were not once embryos.[23] The
whole is just the sum of its parts.

The pervasive influence of this reductive script about persons is manifest in
the focus of the public debate about the ethics of cloning, and of embryo experimentation, on the moral status of the embryo apart from its material and social
context.[24] It was this approach which crucially informed the British House of
Commons ruling in 1990 that experimentation on embryos up to 14 days was permissible because it is in this period that the so-called 'primitive streak' emerges
in the embryo which marks it as 'the beginning of a unique, human individual'.[25]
As Anthony Dyson suggests, this approach reduces the individual person to an
independent 'ethical unit' whereas in reality human beings exist only in interdependence with other human beings.[26] The container narrative fails to account for
an essential aspect of what constitutes human identity, namely its biological character and situatedness in relation to other bodies, to kin, and, Christians would
add, to God, a situatedness which is especially crucial in describing prenatal life.
All humans *were* at one time embryos, and it is only when embryos are removed
from bodies and from natural relations that it becomes possible to regard and treat
them as non-personal, and hence open to acts of volition which include contrac-
tual, legal and scientific procedures and techniques.

In Euro-American culture a person is a hybrid of two essential social traits:
the first is that a person is a unique individual and the second is that a person is
someone who has kin, who exists in a set of relationships with other persons who

---

22   This is also the central argument of 'What is a person?' in Oliver O'Donovan's *Begotten or
Made?* (Oxford: Oxford University Press, 1984): the essays in this book have been formative in a
number of ways for the argument that follows in this chapter.

23   This way of thinking owes its origins to Newton. On the role of the metaphor of container
in Newton's physics, see further Amos Funkenstein, *Theology and the Scientific Imagination in the
Seventeenth Century* (Princeton, NJ: Princeton University Press, 1986).

24   A recent paper by Claire Foster, who writes for the Archbishops' Council of the Church of
England on bioethics, illustrates the continuing influence of this approach: 'Embryo Research:
Some Anglican Perspectives', *Islam and Christian-Muslim Relations* 16 (2005), pp. 285–95.

25   Chair of the Interim Licensing Authority (precursor of the HFEA) as cited in Marilyn
Strathern, 'The Meaning of Assisted Kinship', in Meg Stacey (ed.), *Changing Human Reproduction:
Social Science Perspectives* (London: Sage, 1992), pp. 148–69 (158).

26   Anthony Dyson, 'At Heaven's Command? The Churches, Theology and Experiments on
Embryos', in A. Dyson and J. Harris (eds.), *Experiments on Embryos* (London: Routledge, 1990),
pp. 89–102 (99).

bore, nurtured and formed that person as a person.[27] Neither individual uniqueness nor kinship relations are sufficient to define a person: rather they act symbiotically so that together they constitute what culturally, morally and socially we know as a person. Kinship is rooted in natural relations, or relations of blood, and describes the way in which the individual finds a community of care and respect. Once the natural roots of kinship are removed, then the social and hence moral status of the child or potential child is more at risk. This helps to explain why embryos which (and no longer who) are not in wombs are not subject to the same considerations as embryos in wombs and why the context of the embryo changes the way in which individuals and societies feel about 'it' and behave toward 'it'. The transition from embodied subject of a life to disembodied object and potential life turns the embryo into an artefact and makes it available for instrumental uses. Worse, it puts it at risk of commodification.

The claim of commodification I am making here does not concern the possible sale of embryos, or the possible creation of a market which connects producers and sellers of such embryos, though these things are possible once embryos are no longer in wombs. Rather, I am interested in the broader claim, first enunciated by Ruth Chadwick, that certain novel medical practices contribute to the making of 'a society in which the bodies of persons are regarded as a resource'.[28] Decisions about embryos which are stored in freezers rather than living in wombs treat these potential children as biological resources which do not inalienably belong to their biological parents. Parents are usually consulted about the fate of 'spare' eggs or embryos. Nonetheless the use of such material is also decided by clinicians, judges and laboratory managers. Their judgements include cost-benefit and means-end calculations, as the instrumental use of spare eggs and embryos in recent scientific experiments indicates. This advances the commodification of potential children and not just of parts of persons as the transplant market has done.[29] It thus changes the very meaning of 'children'.

Some may object that since parents do not 'own' their natural-born children then this transit of embryos from belonging to parents to belonging to laboratories is not morally significant. But one needs to ask why parents do not normally use the language of ownership of their offspring, and especially not parents trained in the Christian tradition. It is not that parents abdicate their ownership in some casual fashion; parents *love*, not own, their children. The language of ownership does not even begin to do justice to bonds of kinship. That parents love their children does not mean that they own or disown them; it means instead that they care for them far more than material objects that they merely possess. This relation is mirrored in Trinitarian theology. The Father did not own the Son, and gave the Son freedom to go to a far country and embrace the human condition, but this does not mean

27 Strathern, 'The Meaning of Assisted Kinship', p. 159.
28 Ruth Chadwick, 'The Market for Bodily Parts: Kant and Duties to Oneself', *Journal of Applied Philosophy* 6 (1989) cited in Stephen Wilkinson, 'Commodification Arguments for the Legal Prohibition of Organ Sale', *Health Care Analysis* 8 (2000), p. 191.
29 Wilkinson makes a useful distinction between two kinds of commodification in 'Commodification Arguments'.

that when the Son was in the far country that the Son did not still belong to the Father. Belonging is not ownership. Analogously parents who love their offspring teach them to love; as adults they mostly leave their father and mother and cleave to another but their capacity and willingness to do this is closely connected with the quality of love which they receive from their parents. Unconditional regard is the name often given to the quality of parental love which promotes the formation of children who in turn are able and willing to love freely and well. And because parents love their biological offspring uniquely they do indeed have rights over them which are legally recognized and upheld by the State except in cases of veri-fiable abuse or neglect. Though they are normally consulted over the destruction or experimental use of spare eggs and embryos, the biological originators of this material do not have rights analogous to the rights of parents in relation to their children, and these embryos become in effect the property of others.[30] That there is a 'need' for such embryos is what Ivan Illich would have recognized as an 'arti-ficially created need' created by clinicians, laboratories and research directors.[31] In this broader sense of commodification, the creation of embryos outside wombs contributes to a society in which embryos become resources and, to the extent that embryos are an essential part of what it is to become a child, then we may say that embryos in freezers contribute to the commodification of childhood.

I am claiming that there is a cultural transition where children pass from gift to commodity in societies where such techniques as IVF, embryo pre-selection and embryonic cloning are practised. A further warrant for this claim arises from the subjection of nascent human life to acts of choice and deliberation. Commodities are objects which are open for consumption, to acts of choice; they are a means, increasingly *the* means, by which people maximize their preferences. Where *some* children are born by such acts of deliberative choice and preference maximiza-tion, there is a sense in which all children born in such societies are increasingly seen as being born by acts of clear volition rather than by some more obscure agency which previous generations might have called destiny. This is the central trope of the film *Gattaca* in which children who are naturally born without the aid of embryo pre-selection are regarded as a class of deviant people who may not hold responsible positions in society because they are seen as less reliable people, more prone to disease or deviancy. For children who are born through embryo pre-selection the claim of commodification is greater still. To the extent that chil-dren born by pre-selection are subjected to such acts of choice, which may extend beyond the prevention of genetic traits linked to certain potential diseases to choice of gender or even of genetic traits which may confer, for example, greater intelligence or height, these children have to some extent become artifices, chosen and designed by those who make them.

30   For comment from the Church of England on the linkage between the creation of spare embryos and experimentation on embryos, see the report of the Mission and Public Affairs Council of the Church of England, *Embryo Research: Some Christian Perspectives* (London: Archbishop's Council, 2003), p. 2.

31   Ivan Illich, 'The Institutional Construction of a New Fetish: Human Life', cited in Nancy Scheper-Hughes, 'Commodity Fetishism in Organs Trafficking', in Nancy Scheper-Hughes and Loïc Wacquant (eds.), *Commodifying Bodies* (London: Sage, 2002), pp. 31–62 (49).

The argument that embryonic cloning turns children into artefacts and commodities customized according to clinical or parental requirements has implications not only for therapeutic cloning but for genetic testing, sex selection, and pre-implantation techniques. Alastair Campbell suggests that the Kantian account of the transcendent and intrinsic significance of individual persons, and the Kantian proscription of using persons as means rather than ends, is the crucial philosophical support for this argument.[32] It is, however, by no means clear that Kantianism effectively addresses the moral problems associated with IVF and human embryonic cloning. In its much criticized anthropological 'thinness' the Kantian approach places considerable weight on the concept 'person' and consequently the onus of deliberation in relation to IVF or embryonic cloning moves to the question 'what is a person?', or 'is an embryo a person?', questions which indeed do occupy much ground in secular bioethics literature. However such considerations have, as in the case of the House of Commons debate, only advanced the move towards embryonic experimentation, and they have not prevented the slide towards instrumental and experimental uses of embryos in some other European countries, such as Sweden.

Far stronger grounds for resisting the slide towards instrumentalism and commodification are provided by traditional Catholic teaching on the ethics of life.[33] In a declaration of the Sacred Congregation entitled *Donum Vitae* the Catholic position is enunciated as follows:

> the fruit of human generation, from the first moment of its existence, that is to say from the moment the zygote has formed, demands the unconditional respect that is morally due to the human being in his bodily and spiritual totality.[34]

The clear logic of this position is that to the extent that a potential person is already present in embryonic life then to that potential person duties are owed which are equivalent to the duties owed to more fully developed persons. This respect is owed because

> from the moment of conception, the life of every human being is to be respected in an absolute way because man is the only creature on earth that God has 'wished for himself' and the spiritual soul of each man is 'immediately created' by God; his whole being bears the image of the Creator. Human life is sacred because from its beginning it involves 'the creative action of God' and it remains forever in a special relationship with the Creator, who is its sole end.[35]

On this account the duties owed to embryonic life derive from the relatedness of every human soul to the image of God, and to their destiny to realize the *summum*

---

32   Alastair Campbell, 'What is Wrong with Cloning Humans?, *Journal of Health Services Research Policy* 8.3 (July 2003), p. 192.

33   John Haldane and Patrick Lee, 'Aquinas on Human Ensoulment, Abortion and the Value of Life', *Philosophy* 78 (2003), pp. 255–78.

34   Congregation for the Doctrine of the Faith, *Instruction on Respect for Human Life in Its Origin and on the Dignity of Procreation: Replies to Certain Questions of the Day* (Vatican: Holy See, 1987), chapter 1.1, www.vatican.va/roman_curia/congregations/cfaith/documents/rc_con_cfaith_doc_19870222_respect-for-human-life_en.html.

35   *Instruction on Respect for Human Life*, Introduction, section 5.

bonum, and this includes the duty not to harm them in order that a greater good may come.

Some argue that this approach accords a degree of respect to embryonic and fetal matter which is not accorded to it in nature, as for example in the case of a heavy period containing such matter or in the case of more advanced spontaneous miscarriage. But this is to neglect the crucial differences with respect to intention and volition as between natural miscarriage and the treatment of stored embryos: so long as an embryonic child is in the womb the body may involuntarily discharge that potential person but since there is no agential intention in such an event – since in other words it is the product of chance and nature and not of choice and will – then the event cannot be an occasion of sin.[36] The destiny of the child was not to become a living person but this is a destiny decided not by a moral agent but by that which lies outside the will of its parents. At the heart of the slide from fertility treatment to embryo pre-selection to experimental use of embryonic life is a new geography of embryos. *Embryos which are not in wombs are in the wrong place.* They therefore become subject to acts of deliberate intention and volition. Embryos in wombs may be loved and lost without harm to the souls of those who do the losing but this is not the case for embryos in sterile fluid in Petri dishes. Embryos that are not in wombs are inevitably treated in a different way and judged according to different moral considerations.[37]

It is often argued that technologies which meet the desire for rational control of human reproduction confer a fuller humanity on those who have access to them.[38] Here humanity is proportional to rational control. In the case of IVF the standard account is that its availability advances the liberty of infertile couples by offering them children when they could not naturally have them. But in the majority of cases infertile couples who enter IVF programmes have used contraception, often for many years, to delay giving birth as they pursue other goals, such as career enhancement, before seeking to have children. And paradoxically the availability of IVF only encourages this trend towards late births, even though the body is less fitted to procreation in middle life, and risks increase not only of infertility but also of genetic damage to children born to late mothers.

The central message of the dystopic novels of Huxley and Atwood, of films such as *Gattaca*, and of essays like O'Donovan's *Begotten or Made*, is that while seeming to advance choice, and hence liberty, the over-determination of procreation involves a loss of that liberty which is conferred on persons who are naturally, spontaneously born, and on parents who are gifted with naturally born children. The reason is that when life is predictable, there is a loss of genuine novelty and surprise. And yet it is precisely in the events we do not control, and have not planned, that we are given opportunities for growth in acceptance of

---

36   Haldane and Lee, 'Aquinas on Human Ensoulment', p. 257.

37   The Papal position as stated in *Evangelium Vitae* is that 'the mere probability that a human person is involved would suffice to justify an absolutely clear prohibition of any intervention aimed at killing a human embryo'. Pope John Paul II, *Evangelium Vitae* (Vatican: Holy See, 1995), III, 60.

38   Joseph Fletcher, 'Indicators of Humanhood: A Tentative Profile of Man', *Hastings Center Report* 2 (September–October 1972), pp. 1–4.

and dependence upon the love of God, and of growth in such virtues as fidelity, patience and perseverance.[39] This perception of the vital spiritual significance of circumstance and surprise powerfully informs Jean Pierre de Caussade's account of the 'sacrament of the present moment':[40] the present has no opportunity to become sacramental, or spiritually significant, if every moment is planned in advance. The characteristically modern attempt to erase surprise and to be 'in control', with the aid of technology, of the conception of life also manifests a lack of trust – trust in divine providence, trust between partners and in sexual relations, and trust in the body. This lack of trust represents a psychological and spiritual affliction for, as Stanley Hauerwas suggests, the ability to live life 'out of control' is crucial to the Christian experience of salvation:

> To live out of control as Christians means that we do not assume that our task as Christians is to make history come out right...those who are without control have fewer illusions about what makes this world secure or safe: and they inherently distrust those who say they are going to help through power or violence.[41]

Hauerwas evokes here, as in many of his essays, a tradition of reflection on destiny and fate according to which it is of the essence of mortal life, both in its tragic and redemptive possibilities, that humans, and other creatures, are not ultimately in control of their fate but are instead subject to the exigencies of circumstance, character, chance and context, all of which are transformed in the eschatological horizon of divine salvation. But whereas in an eschatological perspective it is divine grace which 'transforms fate into destiny',[42] in modern society technique increasingly gives individuals the illusion of control over their lives as it demands ever greater conformity to its various systems.

Theologically it may be that the strongest language we can use to describe the status of an embryo which 'lives' outside the womb is that of exile.[43] The sterile waters of the Petri dish are effectively the waters of Babylon, for in these waters nascent human lives are subjected to the imperial purposes of an alien power which chooses whether they will be given the opportunity of life in the womb, sacrificed on the altar of experimental knowledge, or casually flushed away. How does God respond to the weak and defenceless in exile? According to the Hebrew prophets, exile was the punishment the people of Israel had brought upon themselves for their idolatry and their collective abandonment of the righteousness and justice which the law required of them. The 'sins of the fathers' were visited on the generations who were exiled in Babylon, and forced to 'sing the Lord's song in a strange land'. If consumer culture is Babylon, and so stands under judgement,

39   David H. Smith, 'Creation, Preservation and all the Blessings...', *Anglican Theological Review* 81 (1999), pp. 567–88.

40   Jean Pierre de Caussade, *The Sacrament of the Present Moment*, trans. Kitty Muggeridge (London: Collins, 1981).

41   Stanley Hauerwas, *The Peaceable Kingdom* (London: SCM Press, 1984), p. 106.

42   The phrase is from the title of Samuel Wells's *Transforming Fate Into Destiny: The Theological Ethics of Stanley Hauerwas* (Carlisle: Paternoster Press, 1998).

43   I am grateful to Sam Wells for his comments on an earlier version of this paper in which he suggested the appropriateness of the language of exile.

while also visiting judgement on the 'children of men', what of the embryos who are exiled in this strange land on account of the sins of their fathers and mothers? How might embryos be redeemed? Where for them is the land of promise? For Christians the answer is in one sense straightforward enough: Christ came to redeem the people of God from their subjection to alien powers, and the reign of sin and death. Though put to death by those same powers, he 'disarmed the principalities and powers' (Col. 2.15) in his triumphant resurrection, and 'led captivity captive' (Eph. 4.8). How, though, are embryos redeemed and liberated from their captivity in the Petri dish or the freezer by this eschatological horizon? Is not their very existence outside the womb evidence of the continuing sway of the powers over human life? Some Catholic families, at the encouragement of Pope John Paul II, have taken up the practice of adopting embryos, seeking to redeem them from their dehumanizing exile. Cardinal Basil Hume suggested that for most embryos swift destruction is the best release that they can be offered.[44] But such responses are necessarily only partial in that they do not address the larger cultural script which underlies the practices of storing and experimenting upon embryos.

Christian resistance to this larger cultural script will critique the false narration of personhood which underlies the instrumental use of embryos and will acknowledge the limitations of the language of the moral status of persons, embryonic or otherwise. Personhood for Christians is valued because children, and their parents, experience life as blessing and gift from the Creator of life, and as the fruit of their relations of love and of their own bodies. A properly contextual understanding of personhood allows no easy isolation of one aspect of biological and spiritual identity from another. It involves the claim that from the first moments of the fertilized embryo in the womb there is already present the mystery of life in all its fullness, as the Psalmist affirms:

> For you created my inmost being; you knit me together in my mother's womb. I praise you because I am fearfully and wonderfully made... My frame was not hidden from you when I was made in the secret place. When I was woven together in the depths of the earth, your eyes saw my unformed body. All the days ordained for me were written in your book before one of them came to be. (Psalm 139:13–16)

For the Psalmist, conception is the moment when the soul begins because it is at conception, and not in foetal development, that the Psalmist finds the beginning of a relationship between potential selves and the creator God. The physical form of the self is preceded by this relational understanding of the self. This finds powerful echoes among the early church fathers who argued that the foetus in the womb is the object of God's providential care. Thus Athenagoras, in the midst of a condemnation of the widespread pagan practice of abortion, describes the foetus in the womb as 'a created being, and therefore an object of God's care', and hence

---

44   Cardinal Basil Hume argued that the destruction of frozen embryos was the least worst option once they had been created, while he also proposed that the law be changed to disallow the creation, and hence frozen storage, of spare embryos in IVF procedures: 'Statement made by Cardinal Hume in response to questions raised concerning the treatment of frozen embryos' (5 August 1996), www.catholic-ew.org.uk/cn/96/960805a.htm.

those who kill the foetus in the womb, or at birth, are equally 'chargeable with child-murder'.[45] Later in the second century, Tertullian uses similar discourse in describing the ensouled character of the foetus: 'Now we allow that life begins with conception because we contend that the soul also begins from conception; life taking its commencement at the same moment and place that the soul does.'[46] And in a discussion of abortion he says:

> In our case, a murder being once for all forbidden, we may not destroy even the foetus in the womb, while as yet the human being derives blood from the other parts of the body for its sustenance. To hinder a birth is merely a speedier man-killing; nor does it matter whether you take away a life that is born, or destroy one that is coming to birth. That is a man which is going to be one; you have the fruit already in its seed.[47]

Both Athenagoras and Tertullian wrote as apologists of the Christian faith, and used claims for the ensouled status of the foetus and the immorality of abortion in their efforts to distinguish Christian belief and practice from those of pagans. Writing in the fourth century, St John Chrysostom adopts the language of gift and blessing to distinguish Christian from pagan attitudes to prenatal life, refusing even to use the pagan language of abortion as he clearly regards this language as itself misleading, as if it were not murder to end the life of an embryonic or pre-term child intentionally:

> For I have no name to give it, since it does not take off the thing born, but prevent its being born. Why then dost thou abuse the gift of God, and fight with His laws, and follow after what is a curse as if a blessing, and make the chamber of procreation a chamber for murder, and arm the woman that was given for childbearing unto slaughter?[48]

For biblical and patristic writers the primary consideration in relation to the treatment of a nascent human life is the recognition that an emergent child is a gift of God and an ensouled being in which God's creative and inspiriting love is already active. Participation in divine being provides the key to accounts of embryonic human life in the first centuries of the Christian era. These accounts translate into distinctive Christian practices regarding prenatal and newborn children which involve first refusing to end the life of prenatal children intentionally, or even to describe such terminations as abortions, which of course linguistically distinguishes them from other acts of intentional killing. They include the refusal to expose or otherwise kill a child at birth, and the contrasting practice of infant baptism by which the Church welcomed the newborn into full membership of the body of Christ as ensouled beings.

The Gospel paradigm for the different attitude to children displayed in Christian churches, as against pagan society, in the first centuries of the Christian era was the narrative where Jesus welcomed children in the midst of one of

45   Athenagoras, *A Plea for the Christians*, ch. 35.
46   Tertullian, *The Soul*, ch. 27.
47   Tertullian, *Apology*, ch. 9.
48   St John Chrysostom, *Homily on Romans*, 24.

his preaching episodes and suggested that their trusting faith and capacity for dependence was a model of fidelity to God 'for of such is the Kingdom of God' (Mt. 19.14). Artificially assisted reproduction and the cloning of embryos are not manifestations of a Kingdom in which children are given a central place. On the contrary, a society in which children are becoming objects of human making is one where children are increasingly imagined as commodities available to acts of choice and preference maximization. As O'Donovan argues, this new instrumental approach to human life advances a failure to value those elements of created order, providence and spontaneity on which depend 'the reality of a world which we have not made or imagined' and 'which simply confronts us to evoke our love, fear, and worship'.[49] The artificial making of children is not open to God in the way that natural conception is because it replaces divine providence with human control, and because it reduces the opportunity for humans to encounter a world which is divinely created, and redeemed in Christ, rather than humanly made and redeemed by science. Infant baptism by contrast sets the dependence of human beings on the divine creator in the making of souls, and hence persons, at the heart of the life of the church and the family. The sacrament of infant baptism not only marks the transit of a person into membership of the body of Christ, it also acknowledges the dependence of child, congregation and parents on God in that transition for not only is life a gift, but so too is the reincorporation of mortal life into fellowship with the divine.

In conclusion, there is a clear contradiction between the ends towards which the technological exile of the embryo directs human life and the foundational Christian claim that God in Christ, and not human technique, redeems embodied mortal life from sin and death. There is, though, a clear fit between the exile of the embryo, and its commodifying implications for human life, and the larger *telos* of modern neoliberalism which increasingly subjects all life, human and other than human, to the commodity form and to the logic of contract. Thanksgiving for the birth of children and the sacrament of baptism, performed in the context of Eucharistic worship, are the central ways in which Christians continue to enact a different script and to affirm that children are gifts from God and participate in the love of God from conception to their becoming full members of the body of Christ.

---

49   O'Donovan, *Begotten or Made?*, p. 3.

Chapter 6

NEUROSCIENCE AND THE MODIFICATION OF HUMAN BEINGS

D. Gareth Jones

1. *Are We Simply an Expression of Neurotransmitter Levels?*

In a recent paper in the *Journal of Neurophysiology*,[1] the authors asserted that romantic love is a biological urge distinct from sexual arousal. By using functional magnetic resonance imaging (fMRI) they postulated that the neural profile of romantic love is more like a physical craving than an emotional state. In college students who were in the early stages of new love, the active brain areas were regions like the caudate nucleus, which produces the neurotransmitter dopamine, and which appears to be involved when people desire or anticipate a reward. Even more fascinating was the observation that this region was on the opposite side of the brain from another area concerned with physical attractiveness.

But surely romantic love is more than the product of neurotransmitters and neurons. Is not this denigrating all we hold dear? Can we dismiss Shakespeare's sonnets as nothing more than neuroscientific fantasies? And from here, it is but a short step to dismissing even religious aspirations as nothing more than the outpouring of one neurotransmitter or another. People with certain levels of dopamine in region X have religious sentiments; those with lower levels have no need of religious symbolism. It is all a matter of neural organization, which in turn is an expression of genetic predisposition. What we need are neurotransmitters, not prayer!

This will seem like a re-run of the old neural determinism argument in modern guise. But these days its basis is far less hypothetical than was once the case.

2. *Healing the Person through Drugs*

Along similar lines, there is a push to find drugs to improve the functioning of people's brains to combat everything from depression and anxiety disorders, to stress and shyness, and even to forgetfulness and sleepiness. This is the world of what has been described as 'super-Prozacs',[2] where almost every behavioural disorder becomes amenable to drug-based treatment.

Consider the use of Prozac as an illustration of a class of drugs that affect the activity of a particular neural transmitter, serotonin. These drugs generally act in

---

1 A. Aron *et al.*, 'Reward, Motivation, and Emotion Systems Associated with Early-Stage Intense Romantic Love', *Journal of Neurophysiology* 94 (2005), pp. 327–37.
2 A. Caplan, 'Open your Mind', *The Economist* (23 May 2002), pp. 73–75.

the emotional centres of the brain, and so influence emotional responses, making people more outgoing and combating depression and anxiety disorders.[3] What relevance might this have in the lives of Christians?

Consider the following.[4] Jane wanted to die in spite of concerted prayer and counselling; after taking Prozac she began to feel normal and could experience God in ways previously unknown to her. Trevor's life was dominated by severe depression and outbursts of anger; following a number of weeks on Prozac he was transformed and felt just as he had years previously when he became a Christian. For people such as these, medication is a fact of life, allowing them to experience normality, with its highs and lows. This suggests that even people who are deeply committed to God may have no hope of a normal life without this sort of assistance.

For some this is profoundly disturbing. Can a pill do what we normally think only the Holy Spirit can do? Why do drugs work when prayer and counselling may have failed?[5] Indeed, should we conclude that spiritual problems can be more effectively cured by neuroscience and psychology than by traditional spiritual means? Or, at the very least, that the latter should be complemented by neuroscience and psychology?

However we answer these questions, we need to assert that there is an inextricable link between the biochemistry of the brain, who we are as persons, and how we relate to each other and to God.[6] Certain brain regions are associated with certain functions, like goal attainment and anxiety; consequently, emotional and mental health requires appropriate functioning of such brain regions and their connections. There is, to some degree, a neural basis for emotional wellbeing and so when aberrations occur, the appropriate neural systems need to be targeted.

In these terms Prozac may allow an individual to achieve freedom from emotional dysfunction and specific psychological disorders, thereby enhancing that individual's ability to build character.[7] This becomes important for an individual whose actions stem, not from understanding, but from compulsion or as a means of dealing with feelings of guilt or failure. To the extent that the latter emanate from emotional affliction, it is appropriate that they are tackled in whatever way is best, including possibly the use of Prozac-like drugs. Boivin has commented that the usefulness of Prozac results 'not because our experience of God is nothing more than a brain state, but because God created us as physical beings within which the brain serves as the final omnibus of our thoughts and emotions relevant to him'.[8]

---

3   M.J. Boivin, 'Finding God in Prozac or Finding Prozac in God: Preserving a Christian View of the Person amidst a Biopsychological Revolution', *Christian Scholar's Review* 32.2 (2003), pp. 159–76; M. Porter, 'Shadow and Light: Reflections on Life with Psychotropic Medications', *Sojourners* (March–April 1998), p. 25.

4   Boivin, 'Finding God in Prozac'.

5   C.E. Barshinger, L.E. LaRowe and A. Tapia, 'The Gospel according to Prozac: Can a Pill Do What the Holy Spirit Could Not?' *Christianity Today* (14 August 1995), pp. 34–37.

6   Boivin, 'Finding God in Prozac'.

7   Boivin, 'Finding God in Prozac'.

8   Boivin, 'Finding God in Prozac', p. 171.

It appears then that Prozac can have a role in healing emotional afflictions because we are persons who are part of God's physical creation, a creation that is maladaptive and in need of redemption. Prozac provides a limited means of redressing brain systems that have been warped by many forms of maladaptation, and by itself is one means of effecting short-term restoration. Conversely, brain systems may also be redressed at least to some extent by restoring one's relationship to God. It is an awareness of this causation duality that lies at the heart of an appropriate use of drugs like Prozac by the medical profession. An absence of such awareness opens the way to reductionism and excessive medicalization.

The assumption underlying these thoughts is that the use of Prozac and similar drugs is to remedy a pathology within the brain. Its intent is a therapeutic one, because the individual is indeed ill. Hence, a pill may work because it is rectifying an aberrant synaptic function, just as a pill may work to correct a malfunctioning heart beat. Nevertheless, modifications of an individual's brain are not without their dangers, since they leave untouched social, community and spiritual problems. Consequently, long-term solutions need to take account of this broader relational context within which the indivdiual is seeking to live, since these indirectly influence neural and mental functioning.

Consider some further examples. How much depression is normal? While clinical depression is a recognized clinical entity, what are we to make of the low-grade depression with which many people live their whole lives? Is there any virtue in living with it, if it can be obliterated? Similarly, should we treat hyperactivity (attention deficit and hyperactivity disorder, ADHD) in children who are unduly boisterous and easily distracted? Simply posing these questions highlights the uncertainty of where the boundary between the normal and the pathological lies. The boundary is both relative and moveable. The drugs that treat ADHD can also be used to improve normal mental functioning, for example, problem-solving abilities.[9] These drugs are *neurocognitive enhancers*, that prompt us to question what it means to be a person, to be healthy and whole, to do meaningful work, and to value human life in all its imperfection.[10] They increase the medicalization of human life, although in no way do they represent the first forays into this domain.

A Christian stance would be expected to start from the premise that the welfare of the individual is paramount. In light of this, it is pertinent to enquire whether a pharmacological improvement in an individual's behaviour detracts from the responsibility of that individual for his or her own actions, or whether it enhances it. My response is that, if pharmacological support enables personal accomplishments to flourish, it is to be welcomed by Christians. Nevertheless, there are provisos, a major one being that, in the case of children, enormous care is required to protect them, especially perhaps against the ambitions of parents.

9   M.A. Mehta *et al.*, 'Methylphenidate Enhances Working Memory by Modulating Discrete Frontal and Parietal Lobe Regions in the Human Brain', *Journal of Neuroscience* 20 (2000), RC65.

10   M.J. Farah *et al.*, 'Neurocognitive Enhancement: What Can We Do and What Should We Do?' *Nature Reviews. Neuroscience* 5 (2004), pp. 421–25.

Drug-based approaches to mental disorders can be extremely useful or unnervingly distracting. The direction in which we move reflects world-view presuppositions. Viewing the brain within the broader context of environmental, social and spiritual forces emphasizes the significance of human relationships; conversely, viewing it in isolation separates it from these relationships. The former has room for the therapeutic use of psychopharmaceuticals; the latter inclines towards undue dependence upon reductionistic biotechnological approaches.

3. *The Fragility of the Brain*

It has been known for a very long time that damage to certain brain regions may have grotesque ramifications. This is because our brains appear to be far more closely related to the sort of people we are, than are our liver, kidneys or pancreas.

One does not have to look far to encounter well-known and tragic examples of catastrophic changes to individuals following brain damage. Zasetsky was a young soldier who, in 1943, was hit in the head by a bullet during the Russian offensive against the Germans in the Battle of Smolensk. His brain injury affected almost all facets of his existence. Initially, he was unable to perceive anything, as his world had collapsed into fragments and he was unaware of the existence of the right side of his body. He had forgotten how to carry out mundane tasks and also the names of common objects; he became repeatedly lost, and he had become illiterate. In one sense Zasetsky had been killed on that fateful day, because from this point onwards his life became a living nightmare. By his own admission, he no longer considered himself to be a man 'but a shadow, some creature that's fit for nothing'.[11]

Closer to the present day we encounter a builder and heroin addict who was transformed into an obsessive artist following a cerebral haemorrhage.[12] Following damage to his frontal lobes he manically paints, sculpts and writes poetry, activities that were completely foreign to him prior to the haemorrhage.

These pathological examples suggest there is a direct relationship between injury to the brain and aberrant behaviour. Brain injury 'A' is directly associated with behaviour 'a', a relationship that holds regardless of the theological stance or moral values of the injured individual. But does this also apply to a healthy individual with a healthy brain and a normal behavioural repertoire?

While great care must be exercised in arguing from the abnormal to the normal, it is difficult to escape the conclusion that everything from memories, emotions and higher thought processes, to motor functions and sensory awareness have physical correlates in the brain.

---

11   A.R. Luria, *The Man with a Shattered World*, trans L. Solotaroff (Harmondsworth: Penguin Books, 1975).

12   J. Giles, 'Change of Mind', *Nature* 430 (2004), p. 14.

4. *The Brain is Not a Machine*

Underlying these considerations is the dictum that if an individual's brain is changed that individual will be changed, as if the machine at the core of our being had been adjusted. If this is so, what are the implications for our view of ourselves as beings of significance? Are we little more than neural machines writ large? These are understandable concerns, but they are based on a machine type of model that suggests not only vulnerability, but also fixity and rigidity.

I want to argue that this model is misleading, not because it fails to do justice to Christian aspirations, but to scientific ones. The brain is the antithesis of a machine, simply because it is highly responsive to numerous environmental stimuli at all stages of life. The two-way interactions between an individual's brain, and the worlds internal and external to that individual, point to the richness of its multi-dimensional context. In no sense can the brain be isolated from an individual, in the way in which the heart can be; neither could it be replaced by another brain without destroying the integrity of the individual as the person he or she is known to be.

A hypothetical 'brain transplant' is of an entirely different order to that of a kidney, heart or hip. Patients undergoing the latter are restored to health yet retain their distinctive identity. But a new brain would change the person. All the experiences that made her the person she was would have gone; the memories would have gone; the values and interests would have changed. One might even ask who it is we are now having dealings with: is it the original owner of the brain, or the owner of the body? No matter how speculative and absurd this is, it leaves us confused.[13]

For the biologist, kidneys, hearts and hips serve machine-like functions, regardless of their sophistication or, in the case of the heart, symbolic significance. It is for this reason that they can be replaced by machines, either on a temporary or permanent basis. It is difficult to hold out the same prospect for the brain, not because of our impoverished understanding or primitive technology, but because of what the brain is like. It is in this regard that we can argue that our brains represent us, in the sense that our brains, our bodies, and all that we are as individuals, are integrally interrelated. Far from being a machine-based model, this is a *personal* model, one that I find far more amenable to a Christian perspective as well as being far more satisfactory scientifically.

5. *The Plasticity of the Brain*

In developing a personal model of the brain we need to look more closely at the brain itself, and in particular at one of its features, its plasticity (the term used by neuroscientists to denote flexibility and pliability). Once we do this, we realize that what we are as persons is not laid down once and for all in the genome. While it is true that the basic ground plan for an individual's brain is specified in the

---

13   D. Gareth Jones, *Designers of the Future: Who Should Make the Decisions?* (Oxford: Monarch Books, 2005).

genome, the detailed patterns of synaptic connections that link its innumerable neurons (nerve cells) are fashioned by a host of influences throughout life.[14]

The developmental period is characterized by an initial overproduction of neurons, when there is massive competition between them, with only 50 per cent surviving into adulthood. The fascinating aspect of this phenomenon is that it is the external environment that has a crucial role in determining which synaptic connections between neurons persist and, therefore, which neurons survive. For example, the eyes of the newborn must receive visual stimulation from the external world during the early months of life, in order to fine-tune the structure of the visual part of the cerebral cortex.[15] Even more to the point, this stimulation must occur during quite specific *critical periods* of brain development. External inputs have an extensive influence on the brain-in-formation, interacting with internal drivers to mould the end-product that is the adult brain, and hence the adult individual.

It is because of this close interrelationship between the developing brain and extrinsic factors that normal development can readily be disrupted. Changes to the environment, delays to the arrival of sensory impulses during development, result in a different brain organization from what would have been the case under other conditions.[16] Consequently, numerous influences during pregnancy can have devastating consequences for a child's subsequent intelligence and behaviour. The converse also holds: a stimulating environment during pregnancy and for the first few months after birth can increase the complexity of neural organization, in turn altering many facets of the behavioural repertoire of the growing individual.

It is hardly surprising, then, that it is the specific fine-tuning of the synaptic connections in any given brain that contributes substantially to that individual's uniqueness and personhood. This fine-tuning occurs initially over the crucial period from six months gestation to two years postnatal, and then throughout life, as synapses are lost and replaced in all parts of the brain. The possible variation is well nigh infinite, and the connections are modified in response to all the experiences that constitute our lives as individuals. What we are as people emerges from this ongoing dialogue between the neural and genetic material we inherit, the worlds we occupy, and the worlds we ourselves construct.

Extensive plasticity like this makes possible the enormous range of human beings' intellectual abilities and spiritual gifts. It is on account of these neural capabilities that we can be human persons epitomized by responsibility, creativity and responsiveness to each other and to God. Remove neural plasticity and many of the marks of true human personhood disappear. Our biological uniqueness as individuals is integral to our theological uniqueness as persons created by God.

14   D. Gareth Jones, 'The Emergence of Persons', in Malcolm Jeeves (ed.), *From Cells to Souls – and Beyond* (Grand Rapids, MI: Eerdmans, 2004), pp. 11–33.

15   L. Eisenberg, 'Would Cloned Humans Really be Like Sheep?', *New England Journal of Medicine* 340.6 (1999), pp. 471–75.

16   P. Levitt, B. Reinoso and L. Jones, 'The Critical Impact of Early Cellular Environment on Neuronal Development', *Preventive Medicine* 27 (1998), pp. 180–83.

6. *Physical Intrusions into the Brain*

Since our brains are constantly in communication with the worlds of other humans, myriad cultural forces, and a host of spiritual possibilities, they are not isolated islands. They are open to intrusion by others, possibly helpful, possibly unhelpful. Consider the case of *neurosurgery*, where the aim is to rectify the consequences of an injury or pathology. If a haemorrhage or tumour is threatening the life of the individual, surgery would be generally viewed in exactly the same light as surgery in other parts of the body. The brain is a physical organ to be approached in the same manner as any other organ. The aim is to restore the patient to their former healthy state.

But the same cannot be said for *psychosurgery*, where the aim is to modify the behaviour of an individual in the absence of any brain pathology. By cutting nerve tracts or destroying a localized brain region, the goal is to alleviate depression, anxiety and obsessive compulsive states, or to control anger, extreme violence and sexual problems. The latter group is the more controversial since it contains what are frequently regarded as socially deviant conditions, the control of which may reflect the mores of society rather than the state of a person's brain or personality.

Psychosurgery represents a relatively extreme example of intrusions into people's brains, whether of the crude version in vogue in the 1940s and 1950s or of the far more sophisticated contemporary version. The well-known and long-discredited procedure of frontal leucotomy or lobotomy was undertaken to calm agitated, disturbed and aggressive patients. Its success in doing this appears to have been accomplished by changing their personalities in subtle ways, making them less affectionate, and less able to concentrate. Even more distressing was the disappearance of normal social senses and the ability to plan ahead.

There is, of course, nothing surprising about these changes, since they reflect the functions of the brain regions in question. This illustrates one facet of cerebral localization. Destroy brain region 'F' that controls function 'f', and function 'f' will be seriously affected with possible consequences for that individual's personality.

Consequently, human responsibility cannot be viewed in complete isolation of the state of one's brain, since some individuals are predisposed towards undue aggression, say, by brain pathology. However, the relationship between the two does not seem to be an inevitable one. The brain pathology is only one factor in precipitating the violence, which is generally the end-result of a sequence of inter-relating factors.

In order to explore this issue further, consider a relatively new procedure, the grafting of neural tissue into the brains of patients with Parkinson's disease, with the aim of controlling the worst of the motor symptoms. This is an excellent model, since the loss of neurons in this case is extremely well defined and focused. The rationale behind neural grafting is that the donor cells will supply the required neurotransmitter (in this case dopamine) to replace that which was lost by the death of the original neurons.

Up to the present, well over 350 patients with Parkinson's disease world-wide have received implants of fetal neural tissue in their brain. It appears that neurons obtained from human fetuses (aborted at 8–11 weeks gestation) can survive and function in the brains of these patients leading to limited symptomatic improvement.[17] Unfortunately, in no case has there been a full reversal of symptomatology following grafting. It may come as a surprise to realize that non-human tissue has also been used for grafting – Parkinson's patients have received grafts of porcine fetal neural tissue.[18]

The ethical issues raised by these studies are not my concern in the present context.[19] What is of relevance here is the question of whether the introduction of neural grafts into the brain poses a threat to the integrity of those patients as people. Should we be afraid of the intrusion, or can it be viewed as a legitimate therapeutic approach that stands or falls on its own inherent scientific and clinical merits?

It appears that a neuron's significance stems from its functional capabilities and from the connections and circuits of which it is an integral part. If this is the case, the important factors are the neurons, growth factors and transmitters being transplanted, and the brain regions involved.[20] The origin of the neurons may be far less significant, whether from the same person, another person, or very dramatically and surprisingly even from non-human species. When viewed at the cellular level, neurons from rat and human brains are, in some regards, indistinguishable. What does seem to be important is the complexity of their interconnections within the respective brains, suggesting that it is the overall number of neurons and the manner in which the brain is put together that is crucial to humanness or ratness. This in turn points to the significance of the number of environmental contexts within which neurons develop and function: within the brain, linking the brain to the remainder of the individual's body, and the world external to the body. These together appear to determine an individual's ultimate personality.

7. What is Normal?

How do we determine the boundaries of normality and how do we know when these are being transgressed? As we have seen, there is no easy answer, since

17   A. Björklund *et al.*, 'Neural Transplantation for the Treatment of Parkinson's Disease', *Lancet Neurology* 2 (2003), pp. 437–45; E.D. Clarkson, 'Fetal Tissue Transplantation for Patients with Parkinson's Disease: A Database of Published Clinical Results', *Drugs & Aging* 18 (2001), pp. 773–85; P. Hagell and P. Brundin, 'Cell Survival and Clinical Outcome Following Intrastriatal Transplantation in Parkinson Disease', *Journal of Neuropathology and Experimental Neurology* 60 (2001), pp. 741–52.

18   J.M. Schumacher *et al.*, 'Transplantation of Embryonic Porcine Mesencephalic Tissue in Patients with Parkinson's Disease', *Neurology* 54 (2000), pp. 1042–50.

19   See D. Gareth Jones and K. Galvin, 'Neural Grafting in Parkinson's Disease: Scientific and Ethical Ambiguity', in preparation. The ethical issues revolve around the use of material from aborted fetuses, and one's response to the concept of moral complicity. The clinical trials are still considered experimental in nature. This is because, while grafted tissue survives and makes synaptic connections with the host brain, the reversal of symptomatology is less complete than neurologists would like.

20   D. Gareth Jones and S. Sagee, 'Xenotransplantation: Hope or Delusion?' *Biologist* 48 (2001), pp. 129–32.

routinely accepted biological limits are wide and the concept of normality is broad and somewhat tenuous. Even an answer to the question of whether an individual is able to function acceptably in society has limits, since it depends on the nature of the demands placed upon individuals by their positions.

If the aim of therapy is to cure or prevent diseases that hinder someone from functioning within society, there would appear to be a role for the use of brain-modifying drugs to bring those people's capacities within biological limits. The aim here is to utilize neuroscience to bring wholeness to people suffering from recognizable diseased states.

However, even the concept of 'diseased states' is a malleable one. Consequently, any modification of the brain utilizing psychopharmaceuticals or behaviour therapy should be controlled by the welfare of the people concerned, and not simply to enable them to conform to the mores of society. What is more, people live within communities and societies, and help may rest principally on effecting improvements to the social structures around them. Direct assaults upon their brains may be inappropriate. But no matter how we argue these points, perplexing situations remain, where people's actions appear to be unnervingly dominated by the organization of their brains. Consider these examples.

When the brains of depressed subjects who have committed suicide are examined, it is invariably found that there is a reduction in serotonin, one of the brain's neurotransmitters we have encountered previously, a deficiency in which can lead to a predisposition to impulsive and aggressive behaviour. In some extreme situations this may precipitate not only suicidal thoughts, but the likelihood of acting on them.[21]

Studies of depressed patients have shown a variety of abnormalities in their brains, including a decreased volume of one brain region (anterior cingulate) and abnormal activity in a neural circuit involving the anterior cingulate and the amygdala.[22] However, it has been far from clear whether these abnormalities precede or are caused by the depressed state (a common situation with findings of brain abnormalities). One recent study seems to show that depression-like changes in the cingulate-amygdala circuit are present in healthy carriers of a high-risk gene.[23] In other words, these individuals who have a genetic vulnerability to depression demonstrate brain abnormalities even though they do not show signs of clinical depression.

Are these individuals genetically determined to become depressed, or is something else required to precipitate depression? The general opinion is that it is the latter and that something else is required, namely, environmental factors. For instance, carriers of the high-risk gene may never develop depression unless they

21   C. Ezzell, 'Why? The Neuroscience of Suicide', *Scientific American* (February 2003), pp. 44–51.

22   H.S. Mayberg, 'Modulating Dysfunctional Limbic-cortical Circuits in Depression: Towards Development of Brain-based Algorithms for Diagnosis and Optimised Treatment', *British Medical Bulletin* 65 (2003), pp. 193–207.

23   L. Pezawas *et al.*, '5-HTTLPR Polymorphism Impacts Human Cingulate-amygdala Interactions: A Genetic Susceptibility Mechanism for Depression', *Nature Neuroscience* 8 (2005), pp. 828–34.

are exposed to stressful and traumatic events, especially in early life.[24] Additional genes may also contribute to susceptibility to mental illness.

These data can be viewed from different perspectives. One approach is to conclude that individuals with such abnormalities have little hope of escaping depression or suicide attempts. However, most researchers see the situation as more complex than this, since these abnormalities do not exist in isolation, but are found alongside other abnormalities in the brain and also precipitating social factors. It cannot be determined which comes first – the abnormality or the other factors. Are people at risk of becoming depressed normal or abnormal? The answer appears to be that they are normal, and that what we recognize as depression is a result of myriad factors, some biological but the remainder non-biological.

These examples illustrate very clearly the issues at stake. Simple conclusions based on brain features alone are unlikely to be helpful. Decisions concerning the bounds of normality are not decided solely on the basis of biological criteria; they incorporate environmental, social, philosophical and theological considerations as well. However, the answer is not to ignore brain differences, but to accept that the neural data constitute one crucial ingredient in determining which approaches may be of greatest assistance to people in need.[25]

As embodied individuals, we all function with clearly discernible boundaries. We all have limitations (biological and social) and weaknesses. It would be surprising if there are not neural patterns within our brains corresponding to these limitations and weaknesses. But we would be foolhardy to conclude that these patterns constitute grounds for relinquishing responsibility for our actions (any more than 'spiritual' neural patterns would determine the belief systems and faithfulness of Christians). Responsibility and decision-making are core markers of the human person, even though the personal history of some individuals makes responsible decision-making exceedingly difficult to attain.

8. *Protecting Brains and Protecting Embryos*

Anyone who moves between the worlds of genetics and neuroscience must be struck by a dissonance. Enormous amounts of time, intellectual effort and theological angst are expended on the human embryo and genetics, but relatively little on the human brain and neuroscience. And yet so many of the largely theoretical questions that dominate genetics have been the bread and butter of neuroscience for decades. Means of changing people are readily available in neuroscience but only dimly on the horizon in genetics; ways of enhancing people are commonplace in neuroscience, but are no more than the subject of intense futuristic debate in genetics.

One reason for this contrast may lie in the focus on identifiable individuals in neuroscience but on generalized populations in genetics. A second possibility

24   A. Caspi *et al.*, 'Influence of Life Stress on Depression: Moderation by a Polymorphism in the 5-HTT Gene', *Science* 301 (2003), pp. 386–89.
25   Jones, 'The Emergence of Persons'.

follows from this, namely, that with the brain we are working within accepted biological parameters. Improvement in memory, say, is from one level to another, but there is no expectation that the improvement will exceed the powers of every other human being. It will simply be better than it would have been for that particular individual. In contrast, much of the genetics debate seems to presuppose that manipulation of embryos will lead to grand new visions of what is humanly possible.

The speculative nature of the genetics debate skews consideration of certain core concepts, chief among which is the issue of *enhancement*. The direction taken by the genetics argument tends to suggest that to enhance an embryo and future individual is to breach boundaries far beyond anything contemplated by therapy. But what meaning might this notion have in neuroscience? Take the example of memory again. One might ask: What is wrong with improving one's memory when others have better memories? This is not enhancement in a radical sense even though the individual may emerge with a better memory than ever before. It may even be possible to enhance memory by improving the individual's environment and not just by giving psychopharmaceuticals. What objections could there be to this, even if the end-result is enhanced memory for that individual?

A third consideration follows from the previous two, and this is that while neuroscience brings us face to face with our own frailty and limitations, discussions around genetics tend to emphasize the ideal. Neuroscience and therapy go hand in hand, whereas popular interest in genetics revolves around the grandiose. This may be for no more profound reason than that neuroscience has tended to focus on those with neural or mental problems, while the current state of genetics is largely exploratory and hence, to many, intimidating. This would change if the enhancement of personality traits came to occupy a more central role within neuroscience.

Neuroscience may assist us as we approach what we are as individuals with particular personalities, abilities and attributes. If I had had a better nutritional background during my prenatal days and as an infant, I would have been a modified version of my present self. If there had been a different environment, there would have been different elements to my personality. However, what matters is what I make of the individual that I am: the person I know myself to be. There is no ideal 'me'. I can only work with the raw material I have, since this is what makes me the person I am, as a psychological being as well as a person standing before God. There is no way in which I can even begin to see myself as some hypothetical new creature with superlative powers. It is at this juncture that neuroscience discussions diverge so dramatically from genetic ones, and lend themselves to pastorally helpful Christian input. They point to the importance of recognizing that we are free beings, who are to be responsive to the call and direction of God.

For instance, I cannot blame my brain for my being the sort of person I am. If my brain is injured (or damaged by some pathological process) I have to live with that in the same way that I have to live with an injured gastrointestinal tract or leg. In this sense my brain is just another organ. And yet the significance of an injured brain may be far greater for what I am as a person before God. Does this help us in thinking about genetic modification, in the sense that any modification to (let alone genetic or chromosomal aberration of) the embryo I once was, would have had profound repercussions for the person I now am?

9. *Emerging Considerations*

In the light of this discussion my conclusion is that it is far from inevitable that neuroscience will destroy our stature and worth as human beings. It may even enlarge it and may prove far more insightful than generally realized for genetic discussions. In order to make progress on these matters, a number of fundamental considerations emerge as important.

1   Defining normality and abnormality as though they are rigidly defined and opposing categories is misleading. There is a continuum from normality through to abnormality, the border between the two frequently being murky and relative. The problems we encounter are generally in the grey areas between the two. While I have addressed this matter from a biomedical and clinical perspective, one has to enquire whether there is also a theological aspect. For instance, are there theological imperatives for espousing normality (no depression) rather than abnormality (clinical depression), or is there no meaningful difference between the two? While this is a pressing issue for neuroscience, it surfaces repeatedly in the genetics arena. For instance, if it proved possible to modify an embryo with genetic markers for clinical depression, would the imperatives be the same?

2   The notion of enhancement is equally tenuous, at least in neuroscience. Once again the border between normality and enhancement is unclear and shifting. Enhancement for one person is normality for another, on account of the enormous biological variability between individuals. Allied to this is the question of whether there is any difference between enhancement via direct intrusion into the brain (using drugs, implants or surgery) and environmental (indirect) enhancement (using social or educational means). The former has a reductionistic façade, appearing to abstract the brain, and isolate it from the remainder of the individual's body and persona. The mechanistic overtones are palpable, and the dangers all too obvious. And yet the differences between the two approaches may be more apparent than real, since what counts is the nature of the effect upon the individual as a person responsive to God and their world. It is for this reason that the quality of the environment emerges as so important, since our integrity as people is probably threatened far more by environmental pressures than by direct physical or pharmacological assaults upon our brains. Neuroscientific considerations appear to suggest that opposition to enhancement as a principle is misguided and unhelpful, although the irreversibility of genetic enhancement tends to set it apart from the neuroscientific variety. Were society to move in this direction of genetic enhancement, the context for considering neuroscientific enhancement would change.

3   As embodied individuals, all aspects of our mental functioning, including belief systems, attitudes, prejudices and predilections, will have neural substrates. This is not unique to the brain, since the functioning of other body systems also has physical and chemical substrates. Indeed, this constitutes the basis of traditional medical diagnosis. The degree of sophistication may be quite different, but I see no reason to postulate a difference in principle. It is within this context that we need to recognize that the brain is a particularly vulnerable part of our bodies, damage to which has immense repercussions for what we are as persons before God. Consequently, protection of the brain (ours and other people's) should be recognized as an obligation placed upon all, and especially upon God's people. This protection takes many forms, from ensuring adequate nutrition before birth to protection from injury and infection after birth. It also entails making available treatment of the brain in its many guises, including employing psychopharmaceuticals, psychological therapy, and surgery, within a therapeutic-enhancement paradigm.

4   The relationship between the brain and the environment is an intimate one, pointing to the central importance of relationships to human life. We dare not study the brain as though it were an isolated entity; it is an integral part of the individual's body, and through this with the community to which that individual belongs, both the human community and the spiritual one. Connected to this is the psychosomatic unity of people, according to which our brains are influenced by other body systems and by our overall functioning. Nowhere does the two-way relationship between brain and environment emerge more clearly than here.

5   Plasticity of the brain lies at the core of our personhood, because in its absence we would be unable to learn, to respond to each other or to God, and also to exercise our responsibility as those made in the image of God. Unfortunately, this is a concept so foreign to non-neuroscientists that it is overlooked in theological debate. Its relevance to genetics discussions is also missed, since any individuals who may be born following genetic or embryonic modification will have brains characterized by a high degree of plasticity and, therefore, capable of changing and moulding their worlds. They will be no more robots or prisoners of their biology than we are.

Chapter 7

ENSOULMENT REVISED IN RESPONSE TO GENETICS,
NEUROSCIENCE AND OUT-OF-BODY EXPERIENCES

Gordon McPhate

1. *Introduction*

Francis Crick, doyen of molecular biology, had a remarkable career. It began with
the discovery of the structure of DNA, in collaboration with James Watson. It
ended with studies on the neuroscience of the human visual processing system, in
collaboration with Christof Koch. From genes to brains – from DNA to neurons:
such wide-ranging scientific work was motivated by Crick's concern to find
out what makes us uniquely human. Out of that concern came his Astonishing
Hypothesis:

> The Astonishing Hypothesis is that 'You', your joys and your sorrows, your
> memories and your ambitions, your sense of personal identity and free will, are
> in fact no more than the behaviour of a vast assembly of nerve cells and their
> associated molecules.[1]

From this statement, it is clearly evident that Crick argues for a reductionist materialist understanding of human nature. For Crick, the ultimate prize of scientific
endeavour would be to explain consciousness in terms of neuronal circuits and
neurotransmitter molecules. He asserts that free will has been located functionally in or near the anterior cingulate sulcus, next to Brodmann's area 24 of the
cerebral cortex, implying that free will is an illusion.

By contrast, a biblical anthropology offers a distinctive and alternative understanding, centred on the notion of 'soul'. Traditionally, the term 'soul' has been
applied to consideration of human uniqueness; specifically in relation to life-force,
consciousness, free will and reason. The notion of soul in its Hebrew roots is a
complex amalgam of the meanings of *nepesh* (soul), *ruach* (spirit) and *leb* (heart).
The notion of soul in its Greek roots is no less a confused admixture of *psyche*
(mind) and *pneuma* (spirit), with Synoptic, Pauline and Petrine interpretations
which differ considerably. Certainly, the notion of soul contains within it elements
of other attempts to define human uniqueness: mind, spirit, essence, immortal
being, personhood, identity, selfhood. An associated notion is that of image of
God, which after only one mention in the book of Genesis, echoed in the Letter to
the Colossians, and no definition, has had a profound influence on thinking about

---

1 Francis Crick, *The Astonishing Hypothesis: the Scientific Search for the Soul*
(London: Simon & Schuster, 1994), p. 3.

Christian anthropology. The meaning of image of God is wide ranging – including creativity, moral agency, capacity for altruistic love, possession of wisdom and reason, and capacity for language and sophisticated communication.

It is the purpose of this chapter to re-examine the territory of Christian claims about human uniqueness in the light of the accumulating scientific evidence from biology and medicine and individual experience about who we are.

2. Tourette's, Turner's and Psychosomatic Unity

Let me start with a relatively common disorder which in many ways seems to confirm Crick's Astonishing Hypothesis: Tourette's Syndrome. This syndrome falls firmly in the developing field of behavioural genetics. It is primarily a genetic disorder which produces a highly disturbing energetic and uncontrolled clinical presentation, including bizarre movements, noises, rituals, curses, mannerisms and often inappropriate playfulness. Importantly, administration of the neurotransmitter dopamine can also simulate Tourette's Syndrome, indicating that the genetic disorder is linked to a neurochemical imbalance which is the reverse of that seen in Encephalitis Lethargica, and Parkinson's disease.

Tourette's Syndrome, then, has a genetics, a neurochemistry, and a behavioural pattern. It also has a psychology which is best expressed in the words of a 'sufferer' responding to his treatment with haldol:

> Having Tourette's is wild, like being drunk all the while. Being on haldol is dull, makes one square and sober, and neither state is really free. You 'normals' who have the right neurotransmitters in the right places at the right times in your brains, have all feelings, all styles, available all the time – gravity, levity, whatever is appropriate. We Touretters don't: we are forced into levity by our Tourette's and forced into gravity when we take haldol. You are free, you have a natural balance: we must make the best of an artificial balance.[2]

Autonomy is the issue here. As the patient said, neither the Tourette's state nor the treated state is really free. The patient observes and analyses the situation, and firmly identifies with the Touretty state as being natural to him, and also the source of his vivacity. Given the choice, which is denied him, he would choose to be what he actually is – a Touretter. Other patients speak of their constant struggle to remain socially respectable against the obsessive compulsion to use rude and obscene language. This inhibition represents the exertion of a kind of autonomy in a disorder which in other respects implies a reduction or even loss of autonomy. Indeed there is a schizoid quality in being a Touretter.[3]

Tourette's Syndrome has become much better known through the recently discovered likelihood that Mozart suffered from it.[4] Of course, we shall never be sure of the diagnosis *post mortem*. However, musicologists have already begun

---

2   Oliver Sacks, *The Man Who Mistook His Wife for a Hat* (London: Gerald Duckworth, 1985), p. 3.

3   Oliver Sacks, *An Anthropologist on Mars* (New York: Arnold A. Knopf, 1995), pp.80, 97.

4   Benjamin Simkin, *Medical and Musical Byways of Mozartiana* (Santa Barbara, CA: Fithian Press, 2004). This thesis is the subject of the whole book.

to link the condition with the person, and his finest music. What is emerging is the probability that Tourette's Syndrome gave Mozart a competitive advantage in creativity. As Karl Barth has perceived, Mozart was able to celebrate both the positive and the negative *schattenseite* aspects of creation. In doing so, Mozart reckoned with the finitude and the limitations of the creation itself, including human beings.[5]

In terms of living a life and being creative, there is a bond between the 'I' of the person and the 'it' of Tourette's Syndrome. In terms of autonomy, on the other hand, there is a gulf between the 'I' of the person and the 'it' of Tourette's.

In Tourette's Syndrome, we can delineate a complete chain from genes to neurotransmitters to neural pathways to behaviour to psychology. In many ways, the condition is a paradigm for a whole range of genetically determined behavioural traits and conditions, all of which represent some form of restriction on what it means to be fully human. Yet, unexpectedly, even in this condition, both autonomy and creativity are preserved, perhaps even paradoxically enhanced. Thus, the complexity of being human is not fully explicable by a reductionist interpretation; rather, the integration of all levels of biological organization supports an interpretation of psychosomatic unity.

Another common example of genetic modulation of psychology is becoming evident in studies of patients with Turner's Syndrome, who have one X chromosome deleted. These studies have indicated that there is a genetic locus for social recognition, critical for the formation of relationships, including the development of language skills and social cognition. It seems that the important distinction for such patients is whether their only X chromosome is maternally or paternally derived.

The neurocognitive profile of Turner's Syndrome is characteristic: normal verbal ability, normal intelligence, deficits in visio-spatial ability, visio-perceptual ability, mathematical ability, nonverbal memory, and attention span.[6]

Admittedly, single gene defects are less likely to be the subject matter of behavioural genetics than more complex gene clusters which act in concert to determine some aspects of behaviour. Nevertheless, the work of behavioural geneticists has begun and will undoubtedly continue to provide more and more evidence that human beings are a psychosomatic unity. Such work is congruent with the conclusions of the pioneer analytical psychologist Carl Jung who anticipated the link between genes and psychology in these words:[7]

---

5   Karl Barth, *Church Dogmatics*, III/3 (London: Centenary Press, 1960), pp. 298f. See also Jeremy S. Begbie, *Theology, Music and Time* (Cambridge: Cambridge University Press, 2000), pp. 93–97.

6   J.E. Ross *et al.*, 'The Effect of Genetic Differences and Ovarian Failure: Intact Cognitive Function in Adult Women with Premature Ovarian Failure versus Turner Syndrome', *Journal of Clinical Endocrinology and Metabolism* 89.4 (2004), pp. 1817–22; D. Skuse, K. Elgar and E. Morris, 'Quality of Life in Turner Syndrome in Related to Chromosomal Constitution: Implications for Genetic Counselling and Management', *Acta Paediatrica Supplement* 88. 4280 (1999), pp. 110–13; D.H. Skuse *et al.*, 'Evidence from Turner's Syndrome of an X-linked Locus Affecting Cognitive Function', *Nature* 387. 6634 (1997), pp. 705–708.

7   Carl Gustav Jung, *The Archetypes and the Collective Unconscious*, Collected Works, 9 (London: Routledge & Kegan Paul, 1959), para. 136.

> It is in my view a great mistake to suppose that the psyche of a new-born child is a tabula rasa in the sense that there is absolutely nothing in it. In so far as the child is born with a differentiated brain that is predetermined by heredity and therefore individualized, it meets sensory stimuli from the outside not with ANY aptitudes, but with specific ones. These aptitudes can be shown to be inherited instincts and preformed patterns.

Jung goes even further in his notion of the collective unconscious, which proposes that we all share a common impersonal record of our emergent evolutionary history, consisting of primordial images:[8] 'The unconscious is the unwritten history of mankind from time unrecorded.'

If true (and many psychologists now reject Jung's experimental methodology), the implications of these links between our evolutionary history, our family history, our unique genetic code, and our behaviour and psychology are radical. Arguably, we are less free and less unique than we thought we were, and we share much more with one another than we thought we did. In this view, much of what we are is already given, already preconditioned, before our personal narrative begins to write itself – and it writes our story within the framework of constraints already laid down.[9]

3. *Original Sin and a Flawed Human Genome in Evolutionary Perspective*

It is particularly impactful that the DNA molecule might encode a psychology, as well as the proteins which determine the structure of body and brain. Also, the code is not perfect – it is flawed in one way or another in each one of us. Every version of the human genome carries predisposition to disease, encoding a potential or actual pathology. A good example would be the presence of proto-oncogenes, which when activated can initiate tumour development. Another example would be a coding error in a tumour suppressor gene, again triggering cancer from a single transformed cell.[10] For some, inherited pathology implies original sin. For others, it denies the love of God in creation. Here we encounter a clash with traditional biblical theology.

The Augustinian edifice built upon the notion of original sin cannot fare well in an evolutionary perspective, in which creation is still in process and the world is still becoming rather than being. The disobedience of our First Parents, and the descent of humankind from Paradise, will no longer do. In creation-through-evolution sin, evil, suffering and death become inevitable yet tragic conse-quences of the process. The notion of original sin could only be rehabilitated if reinterpreted in the light of the evolutionary ascent of humankind, and the Jesuit geologist Teilhard de Chardin attempts that reinterpretation in these words:[11]

---

8   Carl Gustav Jung, *Psychological Approach to the Dogma of the Trinity*, Collected Works, 11 (London: Routledge & Kegan Paul, 1958), para. 280.

9   Michael Palmer, *Freud and Jung on Religion* (London and New York: Routledge, 1997), pp. 95–101.

10   F. Macdonald and C.H.J. Ford, *Oncogenes and Tumour Suppressor Genes* (Oxford: Bias Scientific Publishers, 1991), pp. 6–15 and 26–29.

11   Pierre Teilhard de Chardin, *Christianity and Evolution* (New York: Harcourt Brace and Co., 1968), p. 40.

> Original sin, taken in its widest sense, is not a malady specific to the Earth, nor
> is it bound up with human generation. It simply symbolizes the inevitable chance
> of evil which accompanies the existence of all participated being. Original Sin is
> the essential reaction of the finite to the creative act.

Thus Teilhard universalizes original sin, and in so doing requires that Christ's
redemption of the fallen universe is also universal. According to Teilhard, we
must acknowledge that the fall is not a single local event, but rather the fall is
spread throughout the whole of universal history.[12] Of course, this controversial
view necessarily implies a considerable lessening of human responsibility for sin,
although the evil of Auschwitz and its equivalents could never be explained or
excused by the evolutionary process.[13]

4. *God's Gift of Image: Potter and Clay; Adoption and Call?*

The reductionist materialist would say that the human genome itself is sufficient
to explain human uniqueness, the variations in genetic code being able to dif-
ferentiate between individuals. The Christian view is that God authenticates the
existence of each and every new human being from the 'beginning' – although
there is much debate about when exactly that 'beginning' is. I suggest that the
donation of the image of God is a creative act of God which authenticates a new
human being from his or her inception. In an evolutionary context, such a creative
act would be understood best by employing a 'potter-and-clay' model of creation.
Here, the imperfect flawed human genome is the clay, and the donation of Image
is the gift of a potential. This gift might be understood in two ways.

First, the gift of the image of God might be expressed as God's 'adoption' of
the unique human genome of a future person. Second, the gift of the image might
be expressed as God's 'call' into existence. The metaphors of 'adoption' and
'call' are consistent with New Testament and Old Testament approaches respectively, but both represent God's selection or election of the individual as a new
and unique creation. Such selection cannot be based upon the merits or demerits
of the genomic 'clay', but rather on God's free elective choice, and would apply
equally to those who will be disabled in some way, and even to those who are
cloned copies of others, as well as those who are the products of the whole range
of reproductive technologies.

God's image represents a potential for the future, into which future God 'calls'
the individual by 'adopting' their unique genome. The call, then, is from the
future, and is teleological in character. God's call to the individual is only one
strand in a universal call to completion, fulfilment and unity to which the whole
evolutionary process in its broadest sense is subject. As Teilhard de Chardin has
said:[14] 'Only a God who is functionally and totally Omega can satisfy us.' In
short, God in this view draws each of us and the whole world into the future, and
the great consummation will occur at the Omega Point where God is all in all.

---

12   John F. Haught, *Deeper than Darwin* (Boulder, CO:Westview Press, 2003), p. 167.
13   Ian Barbour, *Nature, Human Nature, and God* (London: SPCK, 2002), pp. 51–53.
14   Teilhard de Chardin, *Christianity and Evolution*, p. 240.

We cannot define the 'beginning' at which God's call to existence is issued. The choice and selection is God's, and we cannot second-guess the relevant criteria for the staging of such a selection. However, I do not mean to imply Calvinist predestination in God's election; rather that the future viability of the individual is known only in the mind of God.

As an analogy, compare a first-year medical student, a third-year medical student, a final-year medical student and a qualified doctor. Throughout training, a medical student becomes progressively more like a doctor, and less like a member of the general public. Selection for entry to medical school is based on predicted potential for becoming a doctor, but once chosen, the individual has a new status in the community at every stage between selection and graduation. This fits with an adoption model of ensoulment in which the individual is brought into a (new) relationship with God. Accordingly, I argue that we should give each and every embryo the benefit of the doubt from the very earliest stage, without being able to define exactly when the earliest stage is. In the words of Thomas Torrance:[15] 'Every child in the womb has been brothered by the Lord Jesus. In becoming a human being for us, he also became an embryo for the sake of all embryos, and for our Christian understanding of the being, nature and status in God's eyes of the unborn child.'

5. *Image of God as Mutual Beholding*

The notion of the image of God merits further consideration as God's creative gift when adopting or calling into existence. A new building in Euston Road, London, has a unique work of art at its entrance. On the outside of the building a brown humanoid statue stands facing into the glass-panelled entrance hall. On the other side of the glass, and directly opposite the first statue, stands another brown humanoid statue within the entrance hall itself. Two statues locked in mutual observation, the one of the other, separated only by a pane of glass. This artwork speaks to me of relationship and not resemblance. The gift of the *Imago Dei* brings us face to face with God, as it were, and into relationship with God. Of course, this mutual beholding of human being and God is best and supremely realized in the relationship between Jesus Christ and God the Father. As Origen describes it:[16] 'The Son has ever gazed on the abyss of the Father's Being.'

In Jesus Christ we see not only the image of God, but also the likeness of God, and the notion of a journey from image to likeness is suggested as the context in which our human life narratives might be understood. We are initiated on our way by God's gift of the *Imago Dei*, and are thereby called by God into authentic existence. The response to that call is a journey towards God, from image to likeness, a life journey I would like to call ensoulment. Conventionally, ensoulment has been understood as the creative act of transfer of some kind of 'substance' from God to human being at some stage at the beginning of life: soul transfer. In other words, ensoulment as a single event. Instead, I suggest that ensoulment might be

---

15   Thomas F. Torrance, *The Being and Nature of the Unborn Child* (Lenoir: Glen Lorien Books, 2000), p. 4.
16   Origen, *De Principiis*.

better understood as an emergent process, from image to likeness, with its '*telos*' in God. In evolutionary terms, such ensoulment would be regarded as a process of true hominization, from potential to realized potential.

The problem with the notion of 'soul' is that it is too easily objectified as a 'substance' which is 'added' or 'possessed'. The same could be said of the notion of 'image of God'. What is creatively 'given' by God is a potential for response to God's call, and a potential for relationship with God, and a potential to achieve the 'likeness' of God.[17]

What are we trying to say when we use soul language and image of God language? I think we are trying to say that there is more to a human being than simply a body and a brain. I think we are also trying to say that the something 'extra' is central to who we really are, and indeed 'shapes' us into who we are. Aristotle's idea was that the soul is the 'form' of the body, giving shape to a human being both physically and spiritually. Unfortunately, this idea was contaminated by the Platonic idea of an entity which is transferred, added and possessed.[18]

6. *Damasio's Rejection of Cartesian Dualism*

A split between body and soul, or between body and mind, is inherent in Platonic thinking, and was championed by the philosopher Rene Descartes, whose famous first principle of philosophy was '*cogito ergo sum*,' 'I think therefore I am.' Commenting upon this new-found principle he remarked:[19]

> From that I knew that I was a substance, the whole essence or nature of which is to think, and that for its existence there is no need of any place, nor does it depend on any material thing; so that this 'me', that is to say, the soul by which I am what I am, is entirely distinct from body...and even if body were not, the soul would not cease to be what it is.

The neurologist, Antonio Damasio, has vilified this Cartesian disembodiment, this split between mind and brain, implicit in the split between mind and body.[20] Rejection of the material world in general, and the human body more particularly, and sexual activity most specifically, have been consequences of this Cartesian dualism especially within the Christian Church. Concomitant with that rejection has been an embracing of a whole range of spiritualities. The same sort of dualistic split has informed the development of medicine in the West, such that holistic approaches are rejected, and clinical psychology and psychiatry are practised separately from the rest of physical medicine, almost denying the psychological dimension to many diseases.

For Descartes, the mind does not require neuroanatomy, neurophysiology or neurochemistry. It is pure detached reason: Antonio Damasio attacks this posi-

17   David A. Jones, *The Soul of the Embryo* (London and New York: Continuum, 2004).

18   Plato, *Phaedo*.

19   Rene Descartes, *The Philosophical Works of Descartes,* I. (New York: Cambridge University Press, 1970), p. 101.

20   Antonio Damasio, *Descartes' Error* (New York: Putnam Berkley Group, 1994), pp. 165–201.

tion by researching the neural pathways relating to rationality, and demonstrates that these pathways are also employed in the processing of feelings and emotions. Thus there may be a causal link between emotion and reason which would have horrified Descartes.[21] In conclusion to this work, Damasio states:

> The truly embodied mind I envision, however, does not relinquish its most refined levels of operation, those constituting its soul and spirit. That soul and spirit, with all their dignity and human scale, are now complex and unique states of an organism. And this is of course the difficult job: to move the spirit from its nowhere pedestal to a somewhere place, while preserving its dignity and importance – to recognise its humble origin and vulnerability, yet still call upon its guidance.[22]

Damasio is an advocate of psychosomatic unity, and there is much clinical evidence in support of that position. The famous case of Phineas Gage suggests itself immediately. A pleasant well-mannered young man underwent a total personality change after a three-foot iron bar weighing 13 pounds passed through his skull and brain as a consequence of an explosion at work. Frontal lobe cerebral injury meant that '*Gage is no longer Gage*', according to his former workmates. Neuropathology confirms the validity of the total integration of body, brain and mind – in other words, psychosomatic unity. So also does neuropharmacology.

Let us consider the drug Prozac, and its use as an antidepressant. Experience with this drug for 15 years has demonstrated its remarkable capacity for transforming the personality of the patients who take it. Melancholy and anxiety have been supplanted by optimism and confidence to such an extent, in many patients, that moral sensitivity has been dramatically numbed and disinhibited behaviour is evident. The issue here is that maybe reducing mental suffering can dull the conscience! A psychiatrist who works with Prozac describes the loss of self as experienced by a patient:[23]

> He was a good responder. On Prozac, Philip felt better than well, and he hated it. He had been prematurely robbed of his disdain, his hatred, his alienation. His acute episode of depression had been frightening…but the six months of feeling well were hell for Philip. He felt phony – he did not trust himself.

In the construction of the personality, brain and body matter as much as, or even more than, the mind itself. Experience with Prozac strongly affirms our psychosomatic unity as human beings.

7. *Damasio Revisits Spinoza*

The dominance of mind over matter, of mind over body and brain, was likewise attacked by the philosopher Spinoza. For him, the mind of the individual cannot be immortal, for it is inextricably and functionally tied to the mortal body. The

---

21   Damasio, *Descartes' Error*, pp. 242–49.
22   Damasio, *Descartes' Error*, p. 252.
23   Peter D. Kramer, *Listening to Prozac* (New York: Penguin Books, 1994), p. 291.

two are mutually correlated so that the body informs the contents of the mind, and the mind influences the body.[24]

The neurologist Antonio Damasio has expanded on Spinoza's insights about psychosomatic unity, drawing our attention to clinical syndromes in which the assembly of body images in the mind is interrupted. These sensory input images derive either from 'the flesh', as he puts it, or from special sensory probes, and signal the state of activity of specific body parts. From these inputs the brain constructs neural maps of body events, and from these mental images are generated, and out of these images we construct an inner representation of reality in our minds. In other words, the body literally 'informs' the mind.[25]

Damasio reports that in deafferentation syndromes, such as asomatagnosia, the mind is in disarray if the syndrome is partial, whereas the mind or self is suspended if the syndrome is total. In contrast, amputation of a limb produces the phantom limb phenomenon in which an absent limb is 'perceived' to be present by the mind because the afferent nerve pathways are still intact. Once again, psychosomatic unity is affirmed as the best explanation of the nature of what it means to be human.

What, then, of the spiritual dimension 'beyond' the operation of mind? Damasio understands spiritual experiences to be in the domain of the neurobiology of feelings and emotions – which, of course, is also the domain in which he locates part of the processing pathway for reasoning. For Damasio, reason, emotion and spirit are closely linked to one another through shared neuronal circuitry.[26]

Let me recap. Descartes locates the self in rational, conscious thought processes, whereas Damasio (building on Spinoza) locates our self-identity in unconscious feelings and emotions.

Perhaps no disease process points up the division between dualism and psychosomatic unity more than Alzheimer's Dementia. For the dualist, the 'person' is going or has gone, and the situation is an inexplicable outrage, or yet another example of natural evil which begs the theodicy question.

For the advocate of psychosomatic unity, on the other hand, Alzheimer's Dementia is natural, part of the way the world is. In the words of Glenn Weaver:[27]

> If we are not eternal persons by nature and simply return at death to the material elements from which we were created, then the progressive dissolution of our experiences of self-identity in dementia may be quite natural in God's created order. In this case spiritual suffering may most clearly originate in my struggle to accept the finite character of my identity and surrender my autonomous pride.

24 Benedictus de Spinoza, *The Ethics Part II* (New York: Dover Press, 1955).

25 Antonio Damasio, *Looking for Spinoza: Joy, Sorrow and the Feeling Brain* (London: William Heineman, 2003), pp. 187–217, 147–50.

26 Damasio, *Looking for Spinoza*, pp. 147–50; idem, *Descartes' Error*, pp. 180–84.

27 In Malcolm Jeeves (ed.), *From Cells to Souls – and Beyond: Changing Portraits of Human Nature* (Grand Rapids, MI and Cambridge: Eerdmans, 2004), p. 99.

8. *Ensoulment as Process*

As will be evident, I have hesitated to use the term soul as such, and have identified 'ensoulment' as a process rather than as an event; but one that is initiated by the elective creative donation of the image of God, by the God who calls us each into existence by adopting our human genome. If the ensoulment process is understood in the perspective of psychosomatic unity, then biological neurogenesis and psychogenesis are part of the process as we emerge from unconsciousness to self-identifying beings in community with others and before God. In this understanding, our minds are epiphenomena of our brains and bodies, and as argued earlier, a fundamental aspect of the construction of our minds is a constant awareness of embodiment. In the perspective of psychosomatic unity there is emphasis on the vertical axis between genes and persons, passing through various levels of biological organization. This is nature – but we must also consider nurture, and a wider range of inputs, in this matter of ensoulment.

I want to suggest that there are three types of relationship relevant to the nurturing of our development as souls. First, we are ensouled in a specific setting in life, a *Sitz im Leben*, which provides a cultural context as well as a range of opportunities and a range of constraints. Second, we are ensouled in a web of personal relationships of different kinds and levels with other human beings. Third, we are ensouled in a conscious or unconscious relationship to the God with whose image we began (and continue) our journey. The journey of ensoulment is teleological, moving us from simply having the image of God towards achieving the likeness of God. I therefore suggest that ensoulment could be represented as a relational vector of three components (culture, other persons, God) each of which is fundamentally important. As John Macmurray has expressed it:[28]

> It is our vocation, then, not only to unify ourselves and the World around us, not only to hold together the material and the spiritual, and to express them as an undivided whole, reaching out beyond created limits, we are also to unite ourselves and the World to God, and so divinize creation.

But what can ensoulment mean beyond physical death within a psychosomatic unity perspective? The simple answer is nothing. Death is the real and absolute end; a discontinuity at least. It is a matter of speculation as to what might be beyond that discontinuity. For example, could we be reconstituted as ensouled beings, possibly embodied in some way, from the memory of God, who called us into existence in the first place? In other words, a recall, a kind of resurrection. Admittedly, it is paradoxical to consider the God who (I have assumed throughout) is outside of time, as having a memory! By contrast, the out-of-favour dualist perspective presents no problem with the survival of physical death and immortality. Indeed, in this view life after death is natural.

---

28   C. Schwöbel and C.E. Gunton (eds.), *Persons, Divine and Human* (Edinburgh: T & T Clark, 1991), p. 135.

9. *Evidence from Near Death Experiences*

I have a personal problem. Intellectually, and on the basis of the available neuro-scientific evidence, I am persuaded by the psychosomatic unity position. However, I myself was subject to a near death experience in which my mind was temporarily dissociated from my body, such that I observed my own cardiopulmonary resuscitation following cardiac arrest, from above. Thus my experience forces me to give further credence to the dualist position.

Perhaps no one has done more to provide support for the dualist position than the neuroscientist Sir John Eccles, who consolidated the development of Descartes' idea by Sherrington, a neurophysiologist, and Penfield, a neurosurgeon. One problem which fascinated Eccles was the translation of the free will idea to 'do' something into the muscle contractions which caused the planned action to happen. The other problem which intrigued Eccles was why the two cerebral hemispheres were not functionally identical, such that only the dominant hemisphere can talk for both, and seems to be the unique locus of the conscious self, as discovered by the researches of Sperry.[29]

In considering these problems, Eccles developed his radical dualist-interactionist theory of brain and the self-conscious mind.[30] Psychosomatic unity states that body, brain and mind are all contained in the same world. The dualist-interactionist theory states that the body and the brain are in one world, and the self-conscious mind is in another world; but that there is a liaison between the two worlds in regions of the dominant cerebral hemisphere, the so-called liaison brain. For Eccles, the unity of conscious experience is a consequence of the operation of the self-conscious mind, rather than due to the neuronal machinery of the cerebral cortex itself.

In summarizing the evidence in support of the dualist position, I cannot do better than refer to an editorial comment in the medical journal, the *Lancet*, which appeared in 1978 and which relates to the whole range of near death experiences (NDEs):[31]

> Collected accounts volunteered by survivors bear striking similarities. Amongst the experiences many have described are an initial period of distress followed by profound calm and joy; out-of-body experiences with the sense of watching resuscitation events from a distance; the sensation of moving rapidly down a tunnel or along a road, accompanied by a loud buzzing or ringing noise or hearing beautiful music; recognising friends and relatives who have died previously; a rapid review of pleasant incidents from throughout the life as a panoramic play-back; a sense of approaching a border or frontier and being sent back; and being annoyed or disappointed at having to return. Some describe frank transcendent experiences and many state that they will never fear death again. Similar stories

29   C.S. Sherrington, *Man on his Nature* (Cambridge: Cambridge University Press, 1940); W. Penfield, *The Mystery of the Mind* (Princeton, NJ: Princeton University Press, 1975); R.W. Sperry, *Neurosciences Third Study Program* (Cambridge and London: MIT Press, 1974).

30   J.C. Eccles, *The Human Mystery* (Berlin: Springer Verlag, 1978), pp. 214–29; *idem, The Human Psyche* (Berlin: Springer Verlag, 1980), pp. 27–49.

31   Quoted in P. Badham and L. Badham, *Immortality or Extinction?* (London: Macmillan, 1982), p. 71.

have been reported from the victims of accidents, falls, drowning, anaphylaxis, and cardiac and respiratory arrest.

By definition, these are not experiences of death itself, but of dying. As a sometime physiologist myself, I must admit that many of these reports could be explained in terms of the physiology of the dying brain. What cannot be explained is the fact that there have been many reports of NDE subjects obtaining knowledge through their experiences which they could not otherwise have known by any other means. I give one classic example:[32]

> A middle-aged woman, apparently dead from a heart attack, found herself floating out-of-body near the ceiling of the ward and watched doctors and nurses working frantically to save her. She drifted out of a window and around the back of the hospital, where something odd caught her eye – a tennis shoe on a window ledge. Almost as soon as she had revived, she told hospital staff – and the shoe was recovered! The patient was a stranger to the city, seriously ill in bed, wired up to various machines for the whole of her stay in hospital.

A very much smaller group of NDE subjects report a very different kind of unobtainable knowledge: they experience meeting persons of whose prior death they themselves had been ignorant!

Recent much more objective researches on the NDE out-of-body experience (OBE) phenomena have produced astounding conclusions. On the one hand, Olaf Blanke at Geneva University Hospital has succeeded in creating an OBE in a patient by stimulating her right cerebral angular gyrus electrically, while attempting to treat epilepsy.[33] On the other hand, a study of 344 resuscitated patients in the Netherlands, conducted by Pim Van Lommell, found that 18 per cent of the patients reported NDEs which were detailed by the patients in a highly consistent manner, and yet were unrelated to any of the physiological and pharmacological parameters studied.[34]

The neuropsychiatrist Peter Fenwick recently completed a survey of 60 resuscitation patients of whom seven reported NDEs at Southampton General Hospital, each of whom had been temporarily brain-dead. After cardiac arrest their brain rhythms were flat within 11 seconds, and by 18 seconds it would be impossible for their brains to continue to construct a model of the world. Commenting on the results of the study Fenwick concludes:[35]

> There is now convincing evidence to challenge the current theory that consciousness can only exist inside the brain – and if you can have consciousness without associated brain function, that is enormously important for our understanding of the mind.

32   J. Iverson, *In Search of the Dead: A Scientific Investigation of Evidence for Life after Death* (London: BBC Books, 1992), p. 67.

33   O. Blanke, T. Landis, L. Spinelli and M. Seeck, 'Out-of-body Experience and Autoscopy of Neurological Origin', *Brain* 127. 2, (2004), pp. 243–58.

34   P. Van Lommel, R. Van Wees, V. Meyers and I. Elfferich, 'Near-death Experience in Survivors of Cardiac Arrest: A Prospective Study in the Netherlands', *Lancet* 358 (2001), pp. 2039–2045.

35   S. Parnia, D. Waller, R. Yeates and P. Fenwick, 'A Qualitative Study of the Incidence, Features and Aetiology of Near Death Experiences in Cardiac Arrest Survivors', *Resuscitation* 48 (2001), pp. 149–55.

In many of the NDE accounts, embodiment is a significant issue. On the one hand, the perception is of dissociation and of a new freedom to move, free of restrictions. On the other hand, there is the continued awareness of having 'some kind' of body, which is clearly different from the physical body which has been left. In his Letters to the Corinthians, St Paul makes important references to embodiment. In the First Letter, in chapter 15, he speaks of exchanging a physical body for a spiritual body, a mortal body for an immortal body. In the Second Letter, in chapter 12, he describes what may have been the first recorded out-of-body experience.

10. *Conclusion*

In order to make sense of my own personal experience, I seek to achieve an accommodation between the psychosomatic unity of Spinoza and Cartesian dualism. What I suggest is this. We begin as a DNA code, which somehow contains a collective unconscious memory of our evolutionary and social emergence as humankind. God 'adopts' our genome at an undetermined point in time close to our inception as organisms, and we thereby become truly human beings, at the beginning of a process of ensoulment. Our natural embodied state is that of psychosomatic unity, as we continue to emerge as persons in the ensoulment process – such that mind is an epiphenomenon of brain. This is consistent with Spinoza. However (I speculate) at some unspecified point in time, the mind develops the potential to operate independently of brain, realized in certain extreme circumstances, when it experiences extrasensory perception of the real world, and an extracorporeal awareness of embodiment. This is consistent with Descartes, and my own experience! Admittedly, the timing of the ensoulment process has implications for the ethical status of fetuses and young children and others, which cannot be further treated in this essay.

Thus I suggest that ensoulment may be a process in which dualism emerges as the flower from the seed of psychosomatic unity. Ensoulment (in my view) is furthermore fundamentally a vector, which is teleologically directed towards the life of God, following the elective call of God into existence. The ensoulment vector is relational with respect to *Sitz im Leben*, our neighbours, and God, as we move from self-consciousness to community-consciousness; destined as we are for a resurrection body and incorporation within the community of the Body of God. From Image of God to Likeness of God. Future Perfect.

Chapter 8

MEDICINE, SCIENCE AND VIRTUE

Neil Messer

1. *Introduction*

Much attention is given, by both Christian and non-Christian writers, to speculations about 'transhuman' futures. As several chapters in this volume show, such
relatively distant possibilities for human modification can be assessed morally
and theologically in their own right, and such reflection can do valuable work of
various kinds. However, it is worth remembering that far-reaching speculations
about human enhancement and transhumanism also raise more mundane-seeming
theological and ethical questions similar to those raised in the more everyday
world of medical practice in our own time. One such issue is the morality of
experimentation on human beings: if, for example, we wish to develop anti-ageing
technologies that greatly extend human lifespans, those technologies will have to
be tested on human beings. Although some advocates of the technologies seem
dismissive of questions about the morality of such experimentation,[1] the questions
are hardly insignificant. An account of the ethics of clinical research, including
the restraints that should be placed upon such research, therefore acts as a kind
of 'filter' for the more speculative discussions of human enhancement and transhumanism: it may be that some projects should be ruled out from the outset, for
example, because their development would entail clinical research conducted in
unacceptable ways.

This chapter is intended to be a contribution to the discussion of these questions. In particular, I shall explore the relationship between the ethics of clinical
research and the ethics of clinical medicine. However, I shall suggest that the
ways in which this relationship is commonly presented are problematic, and shall
instead work towards a re-location of the discussion in a theological context. It
might be thought that a discussion of the ethics of clinical research should be conducted as far as possible in tradition-neutral terms in the hope that the conclusions
would prove convincing to health professionals, scientists and policy-makers.
However, for reasons that have been well articulated by Alasdair MacIntyre and
others, I doubt whether there is tradition-neutral ground on which such discussions can be conducted.[2] Therefore, my main concern in this chapter is to ask how

---

1   See, e.g., Holger Breithaupt and Caroline Hadley, 'Curing Ageing and the Consequences'
(interview with Aubrey de Grey), *EMBO Reports* 6.3 (2005), pp. 198–201 (199–200).
2   Alasdair MacIntyre, *Whose Justice? Which Rationality?* (London: Duckworth, 1988),
pp. 349–69.

Christians might speak about the issue in a way that is consistent with Christian tradition. A full response to the question of how this theological perspective might be deployed in public and professional forums must await another occasion.[3]

2. *The Problem and a Popular Solution: Clinical Research and Equipoise*

In the modern era, medicine has had a close and reciprocal – if somewhat ambiguous – relationship with the natural sciences. In recent years that relationship has taken the form of an increasing emphasis upon 'evidence-based medicine', by which is meant, among other things, that 'clinical decisions should be based on the best available scientific evidence'.[4] It is obvious that much of this evidence concerns the effects that treatments have on human beings. Evidence-based medicine therefore requires the use of human subjects in scientific experimentation that inevitably involves some level of risk, burden and harm (however slight) with the aim of benefiting patients in the future by improving medical care. A situation in which some human beings are subjected to risk, burden or harm for the benefit of others brings with it particular dangers of exploitation, which is of course why a large ethical literature has grown up around clinical research, and why heavyweight frameworks of ethical regulation exist to police such research in Western countries.

The central moral dilemma of clinical research can be seen most clearly in the case of those clinical trials in which clinical investigators (who are always also healthcare professionals) recruit their own patients as research participants. These cases appear to involve a conflict of roles on the part of the clinical investigator. As clinician, she has an obligation to do the best for each of her patients, including offering them the most effective and appropriate treatments to which she has access. As investigator, she has an obligation to obtain the best possible data about the effectiveness of the treatments. This typically requires a randomized controlled trial (RCT) in which participants are assigned at random to groups receiving different treatments. For example, in a clinical trial designed to assess the effectiveness of a newly developed treatment, one group will receive the new treatment. This group might be compared with another that receives an established treatment for the same condition, with a group that receives an inert placebo, or with both. Where possible, such trials are double-blind (neither investigators nor participants know which group each participant is in) to avoid the distortion of results by psychological factors.

In such a trial, if the investigator recruits participants from among those of her patients who have the condition in question, it appears that she is *not* doing her best for all of her patients. In order to compare the effects of different regimes, she is placing her patient-participants in a situation where some of them will

---

3 For some brief suggestions, see Neil Messer, 'Healthcare Resource Allocation and the "Recovery of Virtue"', *Studies in Christian Ethics* 18.1 (2005), pp. 89–108.

4 F. Davidoff *et al.*, 'Evidence Based Medicine: A New Journal to Help Doctors Identify the Information They Need', *British Medical Journal* 310 (1995), pp. 1085–86, quoted in Tony Hope, 'Evidence Based Medicine and Ethics', *Journal of Medical Ethics* 21.5 (1995), pp. 259–60 (259).

receive a less effective treatment than others. Some are therefore being subjected to some measure of risk or harm (albeit, in many cases, very slight) for the benefit of others: future patients who will benefit from the scientific knowledge gained from the trial.[5]

During the 1970s and 80s, an apparently promising way of resolving this dilemma was developed: the concept of 'clinical equipoise', or genuine uncertainty within the professional community as to which of the different treatments being compared in an RCT offers the greatest therapeutic benefit. If equipoise exists, then an investigator can be considered as doing her best, in good faith, for all of the patients whom she enlists as trial participants, since she is offering each of them a treatment regime which, for all that she and her colleagues know, could be the best for them.

3. *Critiques of Equipoise*

However, the concept of clinical equipoise has been attacked in various ways. One recent critique has been developed by Franklin Miller and Howard Brody, who argue that the concept expresses a fundamental confusion about the ethics of clinical research.[6] Clinical equipoise is an aspect of what they call the 'similarity position': the view that clinical research must be located within the same ethical framework that governs clinical medicine. Against this view, they advocate the 'difference position': that clinical research is a fundamentally different kind of activity from clinical medicine, and must be evaluated by means of a different ethical framework. To argue that research and therapy are different activities, of course, invites the questions whether, and why, both should continue to be done. Miller and Brody make it clear that they believe clinical research to be a valuable and ethically justifiable activity because it has the potential to improve medical care in the future. In other words, they locate its value within what Gerald McKenny and others have called the 'Baconian project': science and technology are valuable because they allow mastery over nature (which, of course, includes human bodies) in order to relieve suffering and maximize individual choice.[7]

---

5   Participants may benefit in various ways from their participation in trials, but this does not resolve the dilemma, which arises because trials are not conducted with the *aim* of benefiting participants.

6   Franklin G. Miller and Howard Brody, 'A Critique of Clinical Equipoise: Therapeutic Misconception in the Ethics of Clinical Trials', *Hastings Center Report* 33.3 (2003), pp. 19–28.

7   Gerald P. McKenny, *To Relieve the Human Condition: Bioethics, Technology and the Body* (Albany, NY: State University of New York Press, 1997), pp. 17–21. Miller and Brody's discussion in 'A Critique of Clinical Equipoise' is mostly concerned with the first of these aims: to relieve suffering, by improving our understanding of diseases and by developing and testing more effective treatments. In this respect (though not in all others) I shall take my cue from them, concentrating mostly on clinical research directed to the treatment of disease. I have argued elsewhere (e.g. Neil G. Messer, 'Human Genetics and the Image of the Triune God', *Science and Christian Belief* 13.2 [2001], pp. 99–111) that the bar of moral justification should at the very least be set higher for technological interventions on the human body directed towards enhancement than for those directed to the treatment of disease. That being the case, we should likewise be more ethically suspicious of clinical research directed to enhancement than of that directed to the treatment of

4. *Testing the Critique: Practices, Virtues and Clinical Research*

While it would be possible to question various aspects of Miller and Brody's argument, their central claim – that clinical medicine and clinical research are fundamentally different activities that must be assessed by means of different ethical frameworks – is an interesting and provocative one that merits serious investigation. One way of exploring this claim (albeit probably not a way that would commend itself to Miller and Brody) is to consider clinical medicine and clinical research as 'practices' in something like Alasdair MacIntyre's sense of the term, to compare the two and to ask how they are related to one another. In what follows, I shall attempt to do so, and shall thereby argue that Miller and Brody are wrong to make a sharp separation between the ethics of medicine and the ethics of clinical research: the two are more closely related than Miller and Brody allow, though the relationship between them is not exactly that described or implied by standard accounts of equipoise.

MacIntyre's definition of a practice is well known:

> any coherent and complex form of socially established human activity through which goods internal to that form of activity are realized in the course of trying to achieve those standards of excellence which are appropriate to, and partially definitive of, that form of activity, with the result that human powers to achieve excellence, and human conceptions of the ends and goods involved, are systematically extended.[8]

It is clear from MacIntyre's account that practices do not lend themselves to simple, one-line definitions. To understand the nature of a particular practice, its internal goods and standards of excellence, one must be part of the community of those who participate in it. Furthermore, that community's understanding of the practice is constantly subject to argument, extension and re-negotiation. If this is so, then the task of comparing medicine and clinical research is likely to be more difficult than Miller and Brody's account would suggest: an adequate account of each will not be supplied by a thumbnail sketch of its aims and goals,[9] but will have to emerge from within the relevant community of practitioners. However, it may be possible to discern the outlines of such an account by thinking about each practice in terms of the *virtues*.

According to MacIntyre, the virtues are qualities that we need in order to achieve the goods internal to practices; some virtues at least, including courage, justice and truthfulness, are essential to sustain any practice. Furthermore, not all qualities that sustain practices can be called virtues, but only those that also contribute to the good of a 'whole human life' and are connected to an 'ongoing tradition'.[10] But it seems likely that any practice will have its own characteristic 'map' of virtues that are particularly important for that practice, or that take

---

disease, so any ethical restraints which we wish to place on the latter are likely to apply *a fortiori* to the former.

8   Alasdair MacIntyre, *After Virtue: A Study in Moral Theory* (London: Duckworth, 2nd edn, 1985), p. 187.

9   See, e.g., Miller and Brody, 'A Critique of Clinical Equipoise', p. 21.

10   MacIntyre, *After Virtue*, pp. 191–92, 275.

particular forms therein. Sketching the 'maps' of virtues characteristic of clinical medicine and of clinical research may enable us to characterize these two practices and their relationship with one another in a more richly textured way than Miller and Brody's rather thin description allows.

a. *Clinical Medicine*[11]

William F. May draws attention to three distinctive 'marks' of the profession of clinical medicine, each of which requires a correlative virtue.[12] First, there is the *intellectual* mark: doctors can be expected to have command of a body of specialist knowledge which guides and informs their practice. Some of this specialist knowledge is scientific knowledge of the sort that could be gained from clinical research. However, the specialist knowledge expected of doctors is not limited to scientific knowledge, since medicine is not merely a technology but also an art. The virtue that is required to sustain the 'healer's art' is *prudence*, in the Aristotelian-Thomist sense of a kind of practical wisdom that can see the patient as a whole, but also as a particular patient, and can discern the action required to care for him or her.

The second mark identified by May is the *moral* mark: while professionals, by definition, make a living from their profession and accordingly have an interest in it, they are not expected simply to be self-interested, but to place their patients' interests before their own. (May, like Miller and Brody, notes a potential conflict here between the priorities of clinical medicine, which place the individual patient's interests first, and those of clinical research, which has other aims than the needs of this particular patient.) To support the trustworthiness that is expected of doctors, the virtue of *fidelity* is required.

Third, there is the *organizational* mark: modern healthcare often demands large resources (and therefore a political and economic system for making those resources available); it takes place in institutions with complex organizational structures, and is delivered by professionals who are members of professional organizations. The practice of medicine is inescapably a corporate activity, and professionals require the virtue that May calls '*public-spiritedness*' to support their co-operation with others in this enterprise. He defines public-spiritedness as 'the art of acting in concert with others for the common good',[13] and it includes

---

11   I begin with medicine for two reasons. First, there is already a well-established body of virtue-ethical reflection on the practice of medicine, but not on the practice of scientific research, and it seems easier and clearer to begin with the better-mapped territory and move from there to the less well charted. Second, and more importantly, even on Miller and Brody's account, medicine provides the context within which clinical research has its value and importance, and without which there would be little need to discuss it.

12   William F. May, 'The Medical Covenant: An Ethics of Obligation or Virtue?', in Gerald P. McKenny and Jonathan R. Sande (eds.), *Theological Analyses of the Clinical Encounter* (Dordrecht: Kluwer, 1994), pp. 29–44. There is obviously much more that could be said about the virtues characteristic of the medical profession, and of other healthcare professions such as nursing. However, within the confines of this chapter, the account derived from May suffices to outline some of the most significant virtues associated with the practices of healthcare and to offer points of contact and comparison with the practice of clinical research.

13   May, 'The Medical Covenant', p. 38.

professional discipline and a concern for justice in the distribution of healthcare resources.

An obvious point, implicit in May's account, is that all of these virtues are located in the context of a practice that is directed to the *good* of suffering people. In other words, the prudence, fidelity and public-spiritedness of medical professionals are all in the service of another virtue, *benevolence*: prudence can be understood as the practical wisdom that helps the doctor to discern in what the patient's true good consists, fidelity as a steadfast commitment to the patient's good, and public-spiritedness as a disposition to act for the common good, recognizing that in any healthcare institution or political community, there are many patients whose good must be considered and sought. As we shall see, this is an obvious point of contact with clinical research, for which benevolence is also said to be central. However, benevolence alone is insufficient to sustain either clinical medicine or (as I shall suggest later) clinical research. Clinical medicine motivated by nothing richer than benevolence would seem to be susceptible to various kinds of distortion. For example, the prudence needed to discern the patient's true good could collapse into mere paternalism or, paradoxically, into an unbalanced 'respect for autonomy' in which acting for the patient's good is reduced to respecting her wishes.[14] Or again, the public-spiritedness that looks to the common good and the wise use of resources could be reduced to a merely utilitarian form of cost-benefit analysis.[15] I shall argue later that if practices motivated by benevolence are to be protected from such distortions, benevolence in its turn must be placed in the context of the virtue of which it is a part, *charity*. This will also entail the claim that the account of the virtues that is needed to sustain the practices of medicine and clinical research, and protect them from the kinds of distortion to which I draw attention, turns out to be a theological one.

b. *Clinical Research*

There are various ways in which one could attempt to define the virtues particularly important in clinical research; the following discussion will proceed by thinking in general terms about the nature of the practice of clinical research and identifying some of the virtues whose absence from all practitioners would render the practice unsustainable.

---

14   The latter is a particular concern in the context of this book, since one feature of speculation about 'transhuman' projects is the way in which technologies initially developed for the treatment of disease come to be seen as tools for the enhancement of human beings in whatever ways those humans' desires or preferences dictate (my thanks to Celia Deane-Drummond and Peter Scott for drawing my attention to this point). In an ethic framed by what McKenny calls the 'Baconian project' (see above, note 7), it becomes difficult, perhaps impossible, to ask whether there are some things that are, and others that are not, *good* for humans to desire or prefer. As we shall see, the virtue approach that I shall propose as a way of framing the ethics of medicine and clinical research suggests a way of asking and answering that question.

15   An obvious example is the widespread use of the quality-adjusted life year (QALY) concept in decisions about healthcare resource allocation; for a critique, see Messer, 'Healthcare Resource Allocation'.

i. Truthfulness

As I observed earlier, MacIntyre holds that truthfulness is one of those virtues that is essential to any practice. It would appear particularly, and distinctively, important for the practice of clinical research. In science generally, it seems that few things disturb the research community so much as scientific fraud and related forms of misconduct, which are typically summarized as 'fabrication, falsification, or plagiarism in proposing, performing, or reviewing research, or in reporting research results'.[16] The popular image is that such misconduct is rare and is severely punished when it does occur.

Fraud illustrates very well MacIntyre's distinction between goods internal and external to a practice.[17] The external goods of scientific research are those (such as career advancement, reputation, power and prestige) that one might hope to gain by doing good research, but that could conceivably sometimes be gained more effectively by falsifying one's results. By contrast, internal goods, such as discovering interesting and valuable things about the world that we did not know before, can only be gained by *doing* good research, not by falsely pretending to. Fraud radically undermines the internal goods of research, and this is particularly evident in clinical research, where the goal is not merely knowledge for knowledge's sake, but an understanding of disease processes and treatments that is directed towards more effective patient care. The virtue of *truthfulness* would seem to be essential for the practice of clinical research.

This simple account of truthfulness as a scientific virtue, though, is open to various kinds of challenge. First, the standard view of scientific fraud – that it is rare, and subject to severe sanctions – is called into question by sociological studies of scientific misconduct. Recent evidence from the United States suggests that, while fabrication, falsification and plagiarism are indeed rare, other forms of questionable scientific behaviour are much more widespread: one-third of the scientists surveyed reported that they had engaged in at least one form of scientific behaviour that, if discovered, would get them into trouble with their institutions or with government regulators.[18] The standard view is challenged more seriously still by radical critic Brian Martin. He has argued that a sharp distinction tends to be maintained within the scientific establishment between fraud, which is held to be a rare aberration and is severely denounced, and other, supposedly less serious, forms of dubious activity, which are accepted or tolerated, because such a distinction serves the interests of those social groups that have the most power over scientific research and the most to gain from it.[19]

---

16   United States Office of Science and Technology Policy, 'Federal Policy on Research Misconduct', www.ostp.gov/html/001207_3.html (accessed 5 August 2005).

17   MacIntyre, *After Virtue*, pp. 188–89.

18   Brian C. Martinson, Melissa S. Anderson and Raymond de Vries, 'Scientists Behaving Badly', *Nature* 435 (2005), pp. 737–38. Examples of such conduct include the failure to present data that would contradict the researcher's own previous research, and altering the design, methodology or results of a piece of research in response to pressure from a sponsor.

19   Brian Martin, 'Scientific Fraud and the Power Structure of Science', *Prometheus* 10.1 (1992), pp. 83–98, www.uow.edu.au/arts/sts/bmartin/pubs/92prom.html (accessed 5 August 2005).

Second, quite apart from specific claims about fraud and misconduct, it is said by Bruno Latour and others that scientific research, when studied 'in the making', is a more untidy human activity than the finished version presented in the journals.[20] Many personal and social factors can filter and sometimes distort what is discovered in scientific research: the perspectives of researchers and the wider social, cultural and economic context in which the research takes place influence what research questions are asked, what data are noticed or ignored and the way the data are interpreted.[21] In clinical research, particular risks of distortion arise from the fact that much of it is commercially sponsored: the interests of commercial sponsors may shape the clinical research agenda in general, influence particular decisions about which trials are or are not conducted, and determine decisions about publication or non-publication of results.[22]

None of this, however, seriously undermines the claim that truthfulness is needed to sustain the internal goods of scientific research. If dubious behaviour *is* more widespread than is usually thought, this simply unsettles any complacent assumption that most scientists successfully exhibit the virtue of truthfulness most of the time. It suggests that truthfulness may be a harder virtue to practise, and the vices opposed to it more subtle and complex, than we tend to suspect. Even if Martin's more radical claim about the relationship between dubious behaviour and power structures is correct, this simply adds two things to the picture: it draws attention to the social and structural, as well as individual, aspects of untruthfulness; and it demonstrates that subtle as well as blatant forms of untruthfulness can be effective means of gaining external goods from research, while at the same time compromising its internal goods.

My claim about truthfulness might seem to be undermined, thirdly, by instrumentalist philosophies of science, which hold that scientific theories cannot be true or false (or, in more modest versions, that it is impossible to know anything about their truth or falsity): theories must be regarded merely as useful instruments for organizing scientific observations and directing future research.

Instrumentalist views are customarily set over against some form of scientific *realism*, which claims *ontologically* that theories can be either true or false, depending on whether they make true claims about the way the world really is, and *epistemologically* that it is possible, at least in principle, to have warranted belief in scientific theories.[23] Jarrett Leplin recently observed that the debate

20   See further, e.g., Bruno Latour, *Science in Action: How to Follow Scientists and Engineers Through Society* (Cambridge, MA: Harvard University Press, 1987).

21   To give just one example, the gender bias of male sociobiologists is said to have had a distorting effect on their observations and theories of human behaviour: see Sarah Blaffer Hrdy, *Mother Nature: A History of Mothers, Infants and Natural Selection* (New York: Pantheon, 1999), pp. 3–117.

22   See, e.g., Trudo Lemmens, 'Piercing the Veil of Corporate Secrecy about Clinical Trials', *Hastings Center Report* 34.5 (2004), pp. 14–18; Carl Elliott, 'Pharma Goes to the Laundry: Public Relations and the Business of Medical Education', *Hastings Center Report* 34.5 (2004), pp. 18–23.

23   Ian Hacking, *Representing and Intervening: Introductory Topics in the Philosophy of Science* (Cambridge: Cambridge University Press, 1983), pp. 28–29.

between these two positions is at an impasse;[24] that being so, this chapter is hardly the occasion to attempt to adjudicate it. It will become clear later that the account of the virtues I am articulating is a teleological one, and this teleology would seem to presuppose some kind of 'realism'. It might, however, turn out to be very different from standard versions, particularly in that it might call into question the characteristically modern separation between ontology and epistemology.[25]

ii. The desire for knowledge

Without a desire for knowledge, clinical research, like other areas of science, would be unsustainable. If clinicians had no desire to understand disease mechanisms better, to seek new treatments that worked better than the existing ones, and so forth, there would be no clinical research. Yet in the Christian tradition, the desire for knowledge is an ambiguous thing. Thomas Aquinas, for example, draws an analogy with physical desires: just as our bodies naturally desire food and sex, so our minds naturally desire knowledge.[26] And like our physical desires, our desire for knowledge can be rightly or wrongly ordered. Thomas calls the rightly ordered desire for knowledge *studiositas*, 'studiousness', and relates it to the cardinal virtue of *temperance* or self-restraint.

The vice of having a disordered desire for knowledge is *curiositas*, 'curiosity', and Thomas says that this disordered desire can take various forms: we can become distracted from the knowledge that we should be pursuing, concentrating instead on 'less profitable' matters; we can seek knowledge from illicit sources; we can seek knowledge of created things without referring that knowledge to God; and we can seek knowledge beyond our capabilities.[27] This typology of *curiositas* offers interesting insights into some of the ways in which medical research might be distorted. A good example of the first type of disorder, being distracted from the more important knowledge by the less profitable, is that the diseases of the wealthy populations of the world attract a disproportionate amount of research effort and funding, and that many diseases of poor populations are neglected by comparison.[28] The reasons are not hard to imagine: the diseases of the wealthy are those most likely to bring commercial research sponsors a return on their investment, are of the greatest concern to the electorates of those governments with the resources to sponsor publicly funded research, and are the ones most likely to worry people wealthy enough to give to medical research charities. This form of *curiositas*, then, is related to the economic distortion of research agendas to which I drew attention under the heading of truthfulness.

24 Jarrett Leplin, 'Realism and Instrumentalism', in W.H. Newton-Smith (ed.), *A Companion to the Philosophy of Science* (Oxford: Blackwell, 2000), pp. 393–401 (393).

25 See John Milbank, 'Knowledge: The Theological Critique of Philosophy in Hamann and Jacobi', in John Milbank, Catherine Pickstock and Graham Ward (eds.), *Radical Orthodoxy: A New Theology* (London: Routledge, 1999), pp. 21–37.

26 Thomas Aquinas, *Summa Theologiae*, 2a2ae.166.

27 Aquinas, *Summa Theologiae*, 2a2ae.167.1.

28 Anon., *DNDi (Drugs for Neglected Diseases Initiative): An Innovative Solution* (Working Draft), www.accessmed-msf.org/upload/ReportsandPublications/19220031120226/DNDi.pdf (accessed 22 September 2005).

It is worth noting that in identifying this type of *curiositas*, Thomas need not be taken to mean that all learning and research should be directed towards topics that are 'useful' in a narrow sense: research that simply enhances our understanding of the created world, without any obvious or immediate practical application, may be worth doing.[29] But where research is aimed directly at meeting human needs, as clinical research is, Thomas's analysis of *curiositas* is a reminder that its priorities must be set by a properly proportioned understanding of the needs that exist. May's medical virtues of prudence and public-spiritedness could help to maintain this proper proportion; this is one way in which a close connection between the practices of clinical medicine and clinical research could help save the latter from some forms of distortion.

Thomas's second type of *curiositas* is to seek knowledge from illicit sources. One way in which this might be done in clinical research would be to use research participants in exploitative or abusive ways; I shall return to this theme later, when discussing benevolence and charity. The third type involves seeking knowledge of created things without referring our knowledge to God. Now the thought that scientific research, if it is to be properly ordered, must make reference to God is unusual in the modern world, to say the least. Indeed, we tend to assume that the extraordinary success and power of science are the result of its *refusal* to refer to God in seeking explanations of natural phenomena, and some scientists and philosophers make much of the stultifying effect that attempts to refer to God are supposed to have had on science in the past.[30] To be sure, it is perfectly possible that inadequate *ways* of making reference to God (for example, treating the Christian doctrine of creation as a rival explanatory hypothesis to neo-Darwinian evolution) can indeed distort the scientific enterprise; it may also be true, as Margaret Atkins argues, that it is not so much the research itself as the character of the researcher that is at risk of distortion by a failure to attend to God.[31] But it is also possible that there are ways in which proper reference to God can safeguard the integrity of the *research* as well as the researcher. Atkins herself points towards one such way when she observes that a defence of scientific research as worth doing, in and of itself, seems to require some notion that the world is both good and ordered, and that this is precisely the understanding articulated by the Christian doctrine of creation.[32] Certainly it is frequently argued that, historically, the Christian doctrine of creation was part of the soil in which modern science took root and flourished in the West.[33]

29   So Margaret Atkins, 'For Gain, for Curiosity or for Edification: Why Do We Teach and Learn?', *Studies in Christian Ethics* 17.1 (2004), pp. 104–17 (111–15).

30   E.g. Daniel C. Dennett, *Darwin's Dangerous Idea: Evolution and the Meanings of Life* (London: Penguin, 1996).

31   Atkins, 'For Gain, for Curiosity or for Edification', p. 111.

32   Atkins, 'For Gain, for Curiosity or for Edification', p. 113. Of course, as her account makes clear, this only amounts to a claim that a Christian understanding of the world as created is *sufficient* to support the attitude that is required in order for science to flourish; it does not (yet) amount to an argument that such a Christian understanding is a *necessary* condition of such an attitude.

33   See, further, Colin E. Gunton, *The Triune Creator: A Historical and Systematic Study* (Grand Rapids, MI: Eerdmans, 1998), pp. 102–16.

The fourth form of *curiositas* described by Thomas is to seek knowledge beyond our capabilities. Again, this may not initially seem like a welcome thought to most scientists, since science seems to have flourished in modern times precisely by *refusing* to accept that there are any problems that will always remain insoluble, or that there are fixed limits to the knowledge of the world of which humans are capable. But at least one form of humility about our knowledge and abilities is familiar to scientists: the honest acknowledgement of the *current* limits to our knowledge and technical capabilities at any given time. Without such an honest admission of how little we know, and how little we are able to do, it would be easy for researchers to attempt to go too far too fast, which could lead to all kinds of failures and false starts, and in the case of clinical research, to unjustifiable treatment of participants.

iii. Benevolence

Doing good to others is central to the standard ethical justification of clinical research: that it will improve patient care in the future. A measure of suspicion may be in order: this justification might sometimes be a cover for other motivations, such as corporate profit, career advancement or even an inordinate desire for scientific knowledge. But there is no reason to doubt that the desire to benefit patients is a genuine motivation for many, perhaps most, clinical researchers.

According to Miller and Brody, clinical research has the 'frankly utilitarian purpose' of seeking the good of future patients by means of experimentation on present research participants. This leads them to insist that it is fundamentally different from clinical medicine in that it does not directly aim for the good of the research participants. The view that clinical research should benefit research participants, they argue, is a widespread misconception; it is prevalent perhaps because it meets the 'psychological needs' of clinical researchers who feel uncomfortable about exposing research participants to risk, burden and harm.[34]

But perhaps what Miller and Brody represent as a kind of psychological weakness on the part of investigators is actually a trace of a virtue without which neither clinical medicine nor clinical research would be sustainable in the long run. According to Thomas, willing and doing good to others are aspects of the supernatural virtue of *charity*, which directs our lives towards love of God and love of neighbour for God's sake.[35] But they are only part of it, and if the virtue of benevolence becomes detached from its wider context in charity, it could be at risk of various forms of distortion: for example, into a 'frankly utilitarian' willingness to use some individuals merely instrumentally for the benefit of others. A disposition to treat one's neighbour instrumentally would seem to be contrary to the virtue of charity, and in the long term, this erosion of charity would be likely to undermine the internal goods of both clinical medicine and clinical research. That is to say, a disposition to use human beings instrumentally could not sustain either medical practice or research that were genuinely directed towards relieving suffering and benefiting the needy, though it could sustain (for example) medical

---

34 Miller and Brody, 'A Critique of Clinical Equipoise', pp. 20–21.
35 Aquinas, *Summa Theologiae*, 2a2ae.23–33.

practice directed towards earning a large salary and clinical research directed towards enhancing the investigator's academic reputation or selling more drugs.

5. *Conclusion: Re-locating Clinical Research in a Theological Context*

My attempt to map the practices of clinical medicine and clinical research in terms of the virtues characteristic of each has suggested that they are more closely related to one another than Miller and Brody allow, and that the most basic connection between them is that they both depend crucially on the virtue of charity as it has in view the needs of sick and suffering people. This suggests a way of characterizing the relationship between them. I have already suggested that Miller and Brody characterize the relationship in terms of the 'Baconian project', but if what I have said about charity is right, then the Baconian project as applied to clinical research will tend to undermine itself. Thanks to a thin conception of the human good that owes much to early utilitarianism, a Baconian approach tends to emphasize some aspects of charity (especially benevolence) but neglect others; I have claimed that benevolence, isolated from its larger context in charity, is in danger of turning into a disposition to seek the welfare of *some* at the expense of *others*.

A related point is made by Allen Verhey: by disparaging the 'speculative sciences' of mediaeval scholasticism, Bacon and his successors cut themselves off from resources that are indispensable for guiding the wise and right use of the 'practical sciences', resources that the latter themselves cannot supply.[36] Setting clinical medicine and clinical research in a framework of the virtues gives us access to those resources and offers a way of understanding what it might mean to use the 'practical sciences' wisely and rightly. What this framework offers is a teleological understanding in which human life (in common with the whole created world) is ordered to an ultimate good. Both clinical medicine and clinical research are morally justified insofar as they are oriented towards that good, and they should be practised in such a way as to contribute to its fulfilment.

But it should be clear from what I have already said that the ultimate good towards which these practices are to be oriented must be understood in theological terms: humans and the world are to be understood as created, loved, redeemed and destined for ultimate fulfilment by the triune God. In this perspective, both clinical medicine and clinical research should be practised in a way that enables our lives to be directed towards that ultimate fulfilment. Clinical research, for example, should be practised so as to enable research participants to display genuine, un-coerced neighbour-love by participating in research, and so as to rule out their merely instrumental use and exploitation by researchers.

I have argued, in short, that to sustain the practices of clinical medicine and clinical research requires a number of virtues – supremely, the virtue of charity – that must be understood theologically. But how are these virtues to be cultivated and supported? In *After Virtue*, MacIntyre famously argued that modern Western

---

36 Allen Verhey, *Reading the Bible in the Strange World of Medicine* (Grand Rapids, MI: Eerdmans, 2003), pp. 151–53.

societies have largely lost the language of virtue from their moral understanding and practice, and that the best hope for the survival and recovery of this moral language is the creation of forms of community in which it can flourish.[37] Thanks to Stanley Hauerwas and others, the thought that Christian churches are called to be such 'communities of character' has become very familiar. Hauerwas is well known for arguing that a community very like the Church is needed to sustain the practice of medicine. This may also be true of the practice of clinical research, if I am right that the latter also depends for its coherence and integrity on virtues that must be understood theologically.[38] To scientists brought up on over-mythologized accounts of past conflicts between science and the Church, this may seem a surprising and unwelcome thought.[39] But, as I observed earlier, it is widely held that the Church and its theological traditions provided the intellectual soil in which modern science took root and flourished in the West. Perhaps it still has a role in fostering not only the intellectual, but also the moral context which science needs. If they are to do this effectively, though, Christian communities might first need to cast a thoroughly critical eye over their own life and practice.[40]

37   MacIntyre, *After Virtue*, p. 263.

38   Stanley Hauerwas, *Suffering Presence: Theological Reflections on Medicine, the Mentally Handicapped and the Church* (Edinburgh: T & T Clark, 1988), pp. 63–83. Much more, of course, would need to be said about ecclesiology in order to give a full account of the role of Christian communities in sustaining these virtues. In particular, some account would be needed of the (sometimes uneasy) relationship between the Church as 'community of disciples' and the Church as institution exercising authority in matters of faith and practice: see, e.g., Avery Dulles, *Models of the Church* (New York: Doubleday, 2nd expanded edn, 1987), pp. 34–46, 204–26.

39   John Hedley Brooke, *Science and Religion: Some Historical Perspectives* (Cambridge: Cambridge University Press, 1991), has argued persuasively that such mythologized conflict narratives fail to do justice to the rich and complex relation between science and religion.

40   An early version of this chapter was tried out on colleagues and students at a research seminar of the Centre for Contemporary and Pastoral Theology, University of Wales, Lampeter; I am grateful to those involved, particularly Dr Simon Oliver, for helpful discussions. I also thank my wife, Dr Janet Messer, for her comments and suggestions, and those involved in the St Deiniol's colloquium for valuable discussion of the ideas developed in this chapter.

Part III

Fabulous Humans

Chapter 9

FORECASTING THE FUTURE: LEGITIMIZING HOPE AND
CALMING FEARS IN THE EMBRYO STEM-CELL DEBATE[*]

Jenny Kitzinger and Clare Williams

1. *Introduction*

The media are a crucial site through which public issues are framed, serving as the focus of intense lobbying and acting as an arena within which policy struggles are defined and played out.[1] They can also have a demonstrable, although not predetermined, impact on how we think.[2] How scientific/medical issues are represented in the media is thus an important area to study when examining the battle both for public opinion and for legislative change.

High-profile crises around issues such AIDS, GM foods, BSE and the MMR vaccine have all attracted scholars to attend to media representations of risk and the processes through which these are constructed.[3] Rapid developments in human genetic research, and the associated media hype, have been the focus of particularly intense scrutiny. Researchers have variously examined journalists' relationships with scientific sources, the accuracy or sensationalism of reporting, and how language, metaphors and imagery may be used to frame issues in particular ways.[4] A theme running implicitly through much of this research is the core

* This work was made possible by the support of the ESRC Centre for Economic and Social Aspects of Genomics (CESAGen) in collaboration with Wellcome Trust funding and an ESRC grant (RES-340250003) to Clare Williams. It is reprinted from Kitzinger, J. and Williams, C., 'Forecasting Science Futures: Legitimising Hope and Calming Fears in the Embryo Stem Cell Debate', *Social Science and Medicine* 61, pp. 731–40, Copyright 2003, with permission from Elsevier.

1   D. Miller, J. Kitzinger and K. Williams, *The Circuit of Mass Communication: Media Strategies, Representation and Audience Reception in the AIDS Crisis* (London: Sage, 1998).

2   C. Condit, *The Meanings of the Gene* (Wisconsin: University of Wisconsin Press, 1999); J. Kitzinger, 'Media Templates: Patterns of Association and the (Re)construction of Meaning over Time', *Media, Culture and Society* 22 (2000), pp. 64–84; A. Petersen, 'Replicating our Bodies, Losing our Selves: News Media Portrayals of Human Cloning in the Wake of Dolly', *Body and Society* 8 (2002), pp. 71–90.

3   G. Philo (ed.), *Message Received* (Harlow: Addison, Wesley and Longman, 1999); J. Lewis and T. Speers, 'Misleading Media Reporting?: The MMR Story', *Nature Reviews Immunology* 3 (2003), pp. 913–18.

4   P. Conrad, 'Use of Expertise: Sources, Quotes and Voice in the Reporting of Genetics in the News', *Public Understanding of Science* 8 (1999), pp. 285–302; A. Hedgecoe, 'Transforming Genes: Metaphors of Information and Language in Modern Genetics', *Sciences as Culture* 8 (1999), pp. 209–29; A. Smart, 'Reporting the Dawn of the Post-genomic Era: Who Wants to Live Forever?', *Sociology of Health and Illness* 25 (2003), pp. 24–49.

question of how the media cover risk predictions: how they address the potential social and ethical consequences and evoke 'bio-fantasies' about the impact of current scientific endeavours.

Debates about developments in biotechnology offer a classic example of the dilemmas in contemporary 'risk society'. As Ulrich Beck has highlighted, science is now seen to produce unprecedented implications which outstrip the ability of 'experts' to predict or control them. The argument is, therefore, that technological developments should be opened up for public criticism. The analysis which follows thus, in part, contributes to debates about the media's role in airing diverse concerns from competing sources and their ability to perform, in Beck's terms, as 'cultural eyes' through which citizens can 'perhaps win back the autonomy of their own judgement'.[5]

This chapter is also located within the growing literature emerging from Science and Technology Studies (STS) on the sociology of expectations and prospective techno-science. Our work is founded on the observation that the implications of any scientific/medical developments are not pre-determined by the technological 'facts', and how people respond to such developments is not pre-ordained. The future of science and technology does not result from a linear, or naturally evolving process but rather, 'the future of science and technology is actively created in the present through contested claims and counterclaims over its potential'.[6] From this perspective, analysts need to explore 'how the future is mobilised in real time to marshal resources, coordinate activities and manage uncertainties'.[7] The media have a crucial role to play in these processes. The significant questions for us, then, are: How is the future constructed in the press and television coverage of stem-cell research? How do proponents of research into this new technology invite media audiences to imagine, believe in and endorse one vision of the consequences rather than another? Who do journalists define as having the authority to comment and how do the scientists involved in such advances seek to gain authority not only over molecules, DNA or genes, but over crystal balls and thus the fears or aspirations of publics, policy-makers and investors?

We explore these questions through a case study of UK media coverage of the stem-cell debate during the year 2000, focusing particularly on embryo stem-cell research. This was a crucial year because it was in August 2000 that a committee of experts chaired by the Chief Medical Officer, Professor Donaldson, published its report reviewing the area of embryo research. The report was launched with a fanfare of publicity explicitly in order to generate wider public debate about the future of stem-cell research before its recommendations were considered by parliament. (Policy-makers, at least in the UK, have reacted to previous controversies

5   U. Beck, *Risk Society: Towards a New Modernity* (London: Sage, 1992).

6   N. Brown, B. Rappert and A. Webster, 'Introducing Contested Futures: From Looking into the Future to Looking at the Future', in N. Brown, B. Rappert and A. Webster (eds.), *Contested Futures: A Sociology of Techno-science* (Aldershot: Ashgate Press, 2000), pp. 3–20.

7   N. Brown and M. Michael, 'An Analysis of Changing Expectations: or "Retrospecting Prospects and Prospecting Retrospects"', *Technology Analysis and Strategic Management* 15 (2003), pp. 3–18.

by trying to be seen to adopt more democratic and transparent procedures which invite public involvement.) One of the principal recommendations of the report was to expand the ways in which embryos could be used to include research aimed at increasing understanding about human disease and disorders and treatments. It suggested that such research should be permitted on embryos created either by IVF or by cell nuclear replacement (CNR) subject to the controls in the 1990 Human Fertilisation and Embryology Act. (CNR involves inserting the nucleus of an adult cell into a donated egg from which the original nucleus has been removed, a process often referred to as cloning.) The Government subsequently drafted regulations to turn these recommendations into law but allowed members of parliament a free vote (not determined by party membership). The Human Fertilisation and Embryology (Research Purposes) Regulations 2001 were passed by the House of Commons on 19 December 2000.

The following section of this chapter presents our research methods. We then present a brief overview of the media coverage before going on to examine the different vocabularies, definitions and frames used by each side in this debate and explore the strategies used to frame utopian hopes as ultimately more credible, valid or worthy than the dystopian fears. Our discussion reflects on the implications of this research for understanding media coverage and the human genetics debate in the context of media studies theory, debates about risk society and the emerging field of the sociology of expectations.

2. *Method*

Our analysis draws on a comprehensive archive of reporting about all aspects of human genetic research for the year 2000 in all national UK newspapers and main TV news bulletins.[8] Each newspaper article in this archive is indexed onto computer by its date, headline, journalist and format (e.g. news report, editorial or feature article). Each TV bulletin is coded in a similar way (adapted to take into account differences in the media form e.g. coding for 'studio-based' and 'outside broadcast'). Every item is also coded for the main story focus, who was quoted and any potential medical, ethical, social or legal risks which are mentioned. Scanning this database for all articles about stem-cell research identified two periods of intense media interest: 13 to 30 August 2000 around the release of the Donaldson report and 19 to 21 December 2000 around the time of the subsequent parliamentary vote. During these two time periods there are a total of 55 newspaper items and eight TV news bulletins about embryo stem-cell research.

An in-depth quantitative and qualitative analysis of this sample was conducted. We examined the conflicting visions of utopia/dystopia presented in the media, paying close attention to metaphors, references to science fiction and use of historical analogies (e.g. the moon landing). We also looked closely at sentence structures and vocabulary. The latter included, for example, analysis of time

---

8   J. Kitzinger, L. Henderson, A. Smart and J. Eldridge, 'Media Coverage of the Social and Ethical Implications of Human Genetic Research'. Final report for The Wellcome Trust, 2003.

references (e.g. 'now', 'soon', 'eventually') and all mentions of 'hope', 'promise' and 'potential'. We also examined the use of words implying *certainty* (e.g. 'guarantee', 'shall'), those implying *possibilities* (e.g. 'expectation', 'prospect', 'chance') and those emphasizing *uncertainty* (e.g. unlikely, doubtful).

In addition to the above we looked at how authority was assigned (or not) to different perspectives by journalists and by their sources. This included analysing how speakers were introduced and how their contributions were framed (e.g. as mainstream or marginal). We also systematically coded all uses of the word 'expert' and associated adjectives (e.g. 'respected', 'eminent', 'leading') as well as terms suggesting collective opinion (e.g. the 'scientific community' or the 'medical world').

As with all analysis of texts it is important continually to reflect on context and not to assume a strict quantitative/qualitative divide.[9] *Qualitative* analysis alone can convey false impressions of the overall pattern of coverage and fail to attend to significant diversity. *Quantitative* analysis on its own can be equally misleading. Content analysis can never be conducted as if it were a merely mechanical process – counting words regardless of how they are being used. We repeatedly returned to the full texts to ensure that we combined qualitative insights with quantitative thoroughness. The word 'potential', for example, is used within the stem-cell debate in very different ways. It can be used to refer to the medical potential of stem-cell research, the biological potentiality of stem cells themselves, or to the potential of an embryo to become a baby. Attention to the different ways in which words are employed is a crucial part of our analysis. In the presentation of our findings we always draw attention not only to patterns in the coverage, but also to any *exceptions*.

Before presenting our findings it is important to note some of the limitations of this research. Ideally media research should take into account the entire circuit of mass communication – including studying how media reports are produced and how they are consumed. However, our main focus here is simply to highlight the different rhetorical techniques used in the public debate, making more transparent some of the ways in which hopes are presented as having greater credibility than fears, highlighting inconsistencies and allowing us to reflect on the issues that might be obscured in the process.

3. Findings: Utopia or Dystopia? The Key Protagonists and the Balance of Media Coverage

The stem-cell debate came across in the media (and, indeed, played out in parliamentary debates) as a dispute between two sharply contrasting perspectives with little room for ambivalence. The supporters of embryo stem-cell research emphasize the positive practical outcomes which might ensue. News footage from the Donaldson report press conference, for example, shows Professor Donaldson declaring that such research could open up 'a new medical frontier with enormous potential' (*Channel 4 Evening News*, 16 August 2000). He asserts

---

9   J. Lewis, 'What Counts in Cultural Studies?', *Media, Culture and Society* 19 (1997), pp. 83–97.

that 'if the expected breakthroughs occur from this research and move forward into treatment we will be able to make a major contribution to the relief of human suffering' (*BBC 6pm News*, 16 August 2000). Some of the subsequent media coverage conjures up visions of a world free from sickness and disease. Stem-cell research could 'herald the start of a medical revolution' (*Sun*, 17 August 2000), usher in 'the dawn of a new frontier' (*Daily Mail*, 17 August 2000) and be the 'key to unlocking a new chapter in medicine' (Anchor, *C4 Evening News*, 19 December 2000).

In contrast, opponents of the Donaldson recommendations envisage a very different future. Embryo research is not only inherently wrong but also sets dangerous precedents for an accelerated abuse of human life and, eventually, for full reproductive cloning. The proposed changes in legislation represent 'a huge leap in the wrong direction for mankind', 'a dangerous and slippery path' which will 'open the floodgates' (Movement against Human Cloning, *Daily Mail*, 16 August 2000; Cardinal Thomas Winning, *Sunday Telegraph*, 20 August 2000; Research director for Life, *Independent*, 19 December 2000).

Systematic analysis of who is quoted in the press and TV reporting shows that the 'pro' and 'anti' lobby are represented by a narrow cast of characters. The pro-Donaldson side is voiced largely by patients, Labour politicians and scientists (e.g. Professor Lord Winston). Their opponents consist mainly of religious representatives, Conservative politicians and anti-abortion activists. Tables 1 and 2 illustrate clearly this stark, binary division.

Analysis of how often each side is quoted, and the amount of airtime devoted to each, shows that the media coverage displays a fairly 'balanced' use of sources. However, not only is there (as we will show) a subtle privileging of the 'pro' position but many journalists also explicitly come out in *favour* of embryo stem-cell research. (This was particularly true in the newspapers, rather than in the TV news bulletins – the latter having an obligation to objectivity.) The Science Editor for the *Express*, for example, robustly supports stem-cell research under the headline, 'Only the Devil could say no to cloning: We have the power to change medicine forever – it's time we used it' (*Express*, 16 August 2000). Such positive sentiments are also reflected in editorials across a wide range of newspapers: 'The benefits of cloning leave no room for doubt' (*Express*, 17 August 2000); 'Say yes to cloning research' (*Financial Times*, 17th August 2000); 'A power for good. Medical gains which must not be shunned' (*Guardian*, 17 August 2000).

Table 1: Newspaper sources: showing number of articles containing quotes from people in these categories *supportive* of or *opposed* to embryo stem-cell research

Categories*	No. of articles containing quotes supportive *of stem-cell research***	No. of articles containing quotes critical *of stem-cell research****
Scientists/doctors	22	0
Labour MPs	10	0
Professor Donaldson	6	0
Govt spokesperson	4	0
Patients/Patients' support groups	4	0
Lib-dem MPs	2	0
Lord Alton (cross-bench peer, Pro-Life campaigner)	0	3
Conservative MPs	1	14
Religious figures	0	13
Prolife groups	0	13
Genetic watchdog groups	0	3
Other	3	2
Totals	52	48

* These categories are based on the introductions provided in the article. For example, Peter Garrett was sometimes introduced as the Director of Life and sometimes as a representative of the 'Movement Against Human Cloning'. Similarly Ann Begg sometimes spoke of her experiences as a patient, but was usually introduced as an MP. Such diverse presentations can be an important lobbying strategy and are discursive acts in themselves. The categories used in coding any individual always echo the media framing rather than superimposing other categories.
**This is the number of articles containing this type of quote, not the number of quotes.
***There were only five quotes which could not easily be assigned to the pro or anti research categories, and these have been excluded from the table.

Table 2: TV news sources: showing number of bulletins featuring people in these categories *supportive* of or *opposed* to embryo stem-cell research

Category of person (as identified on screen)	No. of bulletins featuring this category of person speaking in support *of embryo stem-cell research (and the total time for which they spoke)**	No. of bulletins featuring this category of person speaking against *embryo stem-cell research (and the total time for which they spoke)*
Labour MPs	5 (372 secs)	0
Professor Donaldson	4 (69 secs)	0
Other scientists/doctors (Profs Winston/Higgins)	2 (198 secs)	0
Patients/Patients' support groups	5 (128 secs)	0
Ruth Deech (HFEA)	1 (15 secs)	0
Conservative MPs	0	3 (38 secs)
Pro-Life spokesperson	0	2 (27 secs)
Genetic watchdog groups	0	2 (177 secs)
Helen Watt ('clinical researcher and Roman Catholic')	0	2 (30 secs)
Jacqueline Laing (Professor of Law, London Guildhall University)	0	1 (20 secs)
Dr Nicholson (Editor of the bulletin *Medical Ethics*)	0	1 (180 secs)

| Dr Bruce (Director of Society, Religion and Technology project for Church of Scotland) | 0 | 1 (180 secs) |
| Rounded totals: | 13 mins | 11 mins |

* Note this is the number of bulletins including footage of this type of speaker, not the number of speakers.

4. Describing Stem-cell Research: Contrasting Language and Evocations

The sharp division between those who were 'pro' and those who were 'anti' embryo stem-cell research carries through not only in their explicit assertions but also in more subtle ways in their use of language about particular biological entities or research techniques. For example, both sides agree that CNR produces cloned embryos; however, proponents prefer to use the term 'pre-embryo' or 'blastocyst', whereas opponents of this process are much more likely to emphasize the term '*human* embryo'. The opponents of Donaldson also sometimes describe CNR simply as 'human cloning', implicitly conjuring up visions of full reproductive cloning. By contrast, those supporting the Donaldson report prefer to describe it as 'therapeutic cloning' – a term which brings the idea that this technique can provide therapies into the label of the experimental process itself. Referring to CNR as 'experimental medical research' would not carry the same positive associations. [10]

Another key word in this debate is 'potential' – in this case this is a word frequently used by protagonists from both sides of the debate. However, the concept of potential is applied in very different ways. For opponents of embryo stem-cell research, 'potential' is about the embryo's latent (or not so latent) status as a human being. This, however, is deconstructed by their critics. Prominent pro-Donaldson MP, Ian Gibson, argues, for example, that 'fertilised eggs die or are destroyed by nature, and the embryo has *only* potential for human life' (*Guardian*, 15 August 2000, our emphasis). Roger Highfield, for the *Telegraph*, goes further: 'the potential of an egg fertilised in the lab is limited: it will not develop for more than a few days unless placed in the womb. And the concept of potentiality is too broad to be useful' (*Daily Telegraph*, 16 August 2000). The same deconstruction of the concept of 'potential' is never applied to the potential of stem-cell research itself. Instead, pro-Donaldson protagonists make liberal use of the language of potential (on 54 occasions) to evoke the possibilities opened up by such research.

An idealized future is also brought into the imminent present with the juxtaposition of words such as 'prospect' and 'now' (*Guardian*, 17 August 2000) and the positive use of 'can' and 'could' (employed much more often than more uncertain terms such as 'may' or 'might'). An appeal by an MP that 'Future generations should not *needlessly* be condemned to a "living hell"' is clearly based on an assumption that such suffering is unnecessary because a cure can be found

10   C. Williams, J. Kitzinger and L. Henderson, 'Envisaging the Embryo in Stem Cell Research: Discursive Strategies and Media Reporting of the Ethical Debates', *Sociology of Health and Illness* 25 (2003), pp. 793–814.

(*Guardian*, 20 December 2000, our emphasis). A remark from Christopher Reeve (a famous actor whose spinal cord was severed in an accident) that being able to walk again is 'a motivating vision of what *will* happen' leaves little room for doubt (*C4 Evening News*, 16 August 2000, our emphasis). It is only a question of time. Another pro-Donaldson speaker mobilizes a historical reference point in a classic call to action. This representative of a Parkinson's research interest group is, we are told, 'now certain there will be a cure'. 'The same principle applies as putting a man on the moon,' she declares. 'We say we are going to do it and work backwards from there' (*Independent on Sunday*, 13 August 2000).

Struggles to legitimate contrasting visions of the future also involve very different descriptions of the legislative changes recommended by the Donaldson report. Opponents talk of bridges being crossed, cats let out of bags and previous legislation being turned 'on its head' (Peter Garrett, *BBC1 9pm News*, 16 August 2000; *Daily Mail*, 17 August 2000). By contrast, supporters of Donaldson's recommendations repeatedly assert that the proposed legislative changes do not necessitate 'crossing the Rubicon', and do not raise any 'new issues of principle' (*The Times*, 17 August 2000). Proponents also sometimes play down the novelty of the scientific procedures that might ensue. While opponents of embryo stem-cell research present proposed developments as radical and alarming innovation, proponents, by contrast, imply that it is not very different from anything that has gone before. Professor Robert Winston, for example, declares that 'this type of research has been going on for 20 years' (*BBC2 Newsnight*, 16 August 2000) while other supporters draw parallels with more familiar techniques such as blood donation or skin grafts (Ian Gibson, *C4 News*, 16 August 2000; Science editor, *Express*, 16 August 2000).

There is an interesting inconsistency here. In many contexts proponents of embryo stem-cell research use the language of breakthrough and 'frontier' science to underline its innovative status. In other contexts, however, they sometimes normalize it in an effort to make it appear more familiar and acceptable. Such inconsistency is also subtly illustrated by the different ways in which pro-Donaldson protagonists present the suggestion that the research might lead to growing whole limbs or organs. Sometimes they emphasize that such outcomes are *not* imminent. The editorial in the *Express*, for example, reassures readers that 'no one is suggesting that human embryos should be cloned for reproductive purposes. Even cloning a whole organ, such as a liver or heart, remains a long way off' (17 August 2000). A similarly pro-Donaldson report in the *Independent on Sunday* informs us that: 'We are not talking about Dolly-type technology. It is not about growing a new hand or heart' (13 August 2000). However, on other occasions, the possibility of growing new organs is presented in support of the Donaldson recommendations. For example: 'Lord Robert Winston…is backing the change in the law and said it would not be long before scientists can grow whole organs' (*Sun*, 17 August 2000) or 'Ultimately, whole limbs and transplant organs could be generated. The research holds out hope for people like former Superman star Christopher Reeve' (*Daily Mail*, 17 August 2000).

The above discussion has highlighted how pro- and anti-Donaldson protagonists conjure up competing ideas about the nature of the proposed changes and their implied consequences. However, it is not only a question of who can make the *substance* of their point of view appear most convincing. The issue is also who can present *themselves* as most credible and their point of view as most 'right-thinking'. The following section examines how pro-Donaldson commentators sought to characterize their perspective (often with the support of journalists) as ultimately more legitimate by associating their position with the right sort of people, the right 'ways of knowing' or the right sentiments and values.

5. *Constructing the 'Right' Point of View*

a. *The Voice of Rationality, Truth, Expertise and Progress*
The dispute around stem-cell research is represented as a conflict between rationality and emotion, fact and fiction. Pro-Donaldson protagonists regularly acknowledge the 'upset', 'instinctive unease', 'fear', 'horror' and 'abhorrence' some people might feel about embryo research. However, such acknowledgement is used to underline the fact that the potential benefits are so great that they outweigh such concerns. In some cases 'gut' feelings against using embryos are acknowledged in order to dismiss such reactions as the 'yuk factor' or the product of 'hysteria-mongers' misleading the public about the true nature of the research (*Sunday Express*, 13 August 2000; *Express*, 16 August 2000). Although both sides are described by journalists as 'passionate' about their point of view (e.g. *The Times*, 20 December 2000), more rational, less instinctive, commitments are attributed to the pro stem-cell position. We are informed, for example, that 'The Donaldson report...offers a well-balanced approach to one of the most emotive issues in science policy' (*Financial Times*, 17 August 2000) and that the Human Fertilisation and Embryology Authority which supported the report is 'an august and highly rational body' (*Express*, 16 August 2000).

Whilst proponents are basing their conclusions on 'the facts', opponents, it is suggested, are being misled by science-fiction fantasies. When we systematically analysed every reference to science-fiction scenarios, an unexpected finding was revealed. Explicit references to science fiction are not used by opponents of embryo research, but appear instead only when *attributed* to them by *proponents* of the research. The Minister for Public Health, for example, declares, 'claims that these regulations will lead to human reproductive cloning are based in science fiction, not in law. It would be dreadful if fears of science fiction are to prevent research which promises hope to real lives' (*Channel 4 News*, 19 December 2000). These sentiments are echoed by journalists' statements such as, 'Pro-life campaigners and religious groups will denounce the move as a step towards "Frankenstein" technology' or that 'scaremongers' might spread 'unfounded Frankenstein fears' (*Sunday Express*, 13 August 2000; *Express*, 14 August 2000; *Express*, 17 August 2000). Science fiction is thus not so much a way of promoting concern about science, used by the anti-Donaldson camp. Rather it is here used, by Donaldson *supporters*, as a rhetorical weapon to discredit the opposition.

Alongside such descriptions, supporters of embryo stem-cell research make strong claims on the concept of expertise. These claims are routinely supported by journalists. In fact the word 'expert' is used by journalists *exclusively* to describe scientists and medical practitioners who support the Donaldson report recommendations. It is *never* used about ethicists, religious leaders or, indeed, qualified medical practitioners or scientists who oppose them. In addition journalists emphasize the number and range of experts in favour of embryo stem-cell research. The Donaldson press conference was, of course, carefully choreographed to underline the massive support for its recommendations from a range of key scientific and medical bodies. There was also intensive lobbying throughout the period leading up to the parliamentary vote from bodies such as the British Medical Association. This high level of national expert consensus is reflected in journalists' coverage. TV bulletins and newspaper articles list the impressive phalanx of organizations that supported the Donaldson proposals and campaigners often underline the fact that the majority of experts (or all who 'count') are on their side. 'Every *respectable* biologist' would support the Donaldson report, pronounces Professor Winston (*BBC2 Newsnight*, 16 August 2000, our emphasis). Journalists reiterate this point, sometimes implying unanimous support of 'the scientific and medical community' (*Express*, 17 August 2000) or what one journalist calls 'the scientific and medical world' (*Daily Mail*, 17 August 2000).

There are only two exceptions to this type of formulation in the press. An article in the *Daily Mail* (21 December 2000), by their Foreign Service in Berlin, reports on German criticism of the UK parliamentary vote. This includes a statement from the German Chancellor declaring that the 'German research community' is opposed to ending the ban on embryo stem-cell research until the full potential of adult cells is better understood. Such German dissension is also mentioned in a piece by the anti-Donaldson campaigner, David Alton (*Sunday Express*, 13 August 2000).

At the same time as claiming a monopoly on rationality, realism and expertise, proponents of embryo stem-cell research characterize their opponents as old-fashioned, or even Luddite. Ian Gibson, writing in the *Guardian*, epitomizes part of this approach in his piece entitled: 'Already battle lines are being drawn in the stem cell debate. I'm on the side of reason and progress' (15 August 2000). The *Sunday Express* editorial frames resistance to stem-cell research in terms of a traditional 'fear of almost any technological development ... That fear is at its most hysterical when it comes to breakthroughs in the field of genetics' (13 August 2000). Those who seek to stand in the way of medical progress are characterized as anti-science and ultimately anti-democratic, seeking to impose judgements coming from the very margins of contemporary society. The *Express* dismisses concerns as '*medieval* objections' (16 August 2000, our emphasis) and the *Financial Times*' editorial advises that, 'A modern secular society cannot accept the extreme view of some religious and anti-abortion groups' (17 August 2000).

The ways in which 'masculine' authority, boldness, frontier spirit and rationality are contrasted with the implicitly 'feminine' marginal, fearful, hysterical or anti-science voice in the above analysis has been well documented by previous commentators on science.[11] What is particularly interesting about the stem-cell

---

11   S. Harding and J. O'Barr (eds.), *Sex and Scientific Inquiry* (Chicago: Chicago University Press, 1987).

debate, however, is the way in which such traditional ways of asserting legitimacy are combined with more 'emotive' discourses about social and family networks and explicit appeals to compassion and hope.

b. *The Voice of Compassion and Hope*

Although opponents attempt to claim the moral and ethical high ground in relation to human life, supporters of embryo stem-cell research refuse to be cowed by these claims. Instead they make counterclaims by inverting the argument. Challenging 'pro-life' positions, the Wellcome Trust Director declares, 'stem cell research is about life – the potential to cure or create healthier lives for the diabetic, the leukaemia patient, the Parkinson's sufferer or stroke victim' (*The Times*, 20 December 2000). Far from being unethical to conduct such research, it would be 'unethical' not to proceed (*Daily Mail*, 17 August 2000), a 'blinkered approach to this technology is an appalling betrayal of the millions of people who suffer from degenerative diseases' (*Express*, 17 August 2000).

The potential benefits of embryo stem-cell research are not just presented in abstract in terms of the 'millions' to be cured but are seasoned with vivid, personalized vignettes. For example we are told that voting for the Donaldson recommendations is a vote in favour of 'the woman with Parkinson's who struggles with speech so she cannot sing nursery rhymes to her children. The grandfather who cannot enjoy his grandchildren growing up because of the devastation of stroke' (Public Health Minister, *Daily Mail*, 20 December 2000). In addition to such vignettes, the voices of those speaking out about their own, or their families', suffering has a key part to play in the debate (and such personal accounts can have a huge impact on audiences; see Henderson and Kitzinger 1999).[12] For example, Anne Campbell, Labour MP for Cambridge, is 'one of several MPs with heart-rending family stories to share' (*Guardian*, 20 December 2000). Ordinary 'sufferers' are also given great prominence. Two TV bulletins about the Donaldson report open with interviews with people with diseases, in each case giving them precedence over shots of the press conference itself (*BBC1 9pm News*; *ITN Evening News*, 16 August 2000). The importance attributed to the 'patient's voice' is underscored by the fact that, in contravention of traditional news room practices, almost twice as much airtime is given to ordinary people representing patients than to Professor Donaldson himself (see Table 2).

The concept of 'hope' is the last, but by no means least, significant of the terms we wish to highlight in this analysis. Hope is a crucial commodity in this debate. The word 'hope' in relation to embryo stem cells is used 24 times in the 55 newspaper articles and 17 times within the eight news bulletins in our sample. 'Hope' performs specific rhetorical purposes. Media consumers are told that embryo research offers 'hopes of amazing breakthroughs in medicine' (*Guardian*, 18 August 2000) or, more specifically, that 'the hope is that they [stem cells] can be turned into skin, bone, and even one day complete organs, all of them exact matches of the patients' (Neil Dickson, *BBC Evening News*, 19 December

---

12   L. Henderson and J. Kitzinger, 'The Human Drama of Genetics: "Hard" and "Soft" Media Representations of Inherited Breast Cancer', *Sociology of Health and Illness* 21 (1999), pp. 560–78.

2000). The embryo is re-inscribed in this debate. It ceases to represent itself (or the hope of pregnancy) and becomes instead a beacon of hope for the sick. A close-up image of an embryo looms into the foreground on a news bulletin as we are informed that 'those small clusters of cells…could offer hope to thousands living with devastating diseases' (*BBC Evening News*, 19 December 2000). A photograph in the *Mirror* newspaper is simply captioned 'HOPE: an embryo' (20 December 2000).

The use of the word 'hope' means that scientists who might not confidently 'predict' effective treatments are able to invoke limitless, and imminent, potential, while at the same time allowing for an escape clause. Hope is not identified here as a human aspiration or emotion. Nor is there much discussion of the potential gap between wishful thinking and reality. Instead hope is presented as a basis for claims-making and as an imperative to action. The concept of hope substitutes for any serious engagement with the 'ifs' raised in Professor Donaldson's careful statement at the press conference quoted at the beginning of this article ('*if* breakthroughs occur and *if* they translate into treatment'). On the few occasions when pro-Donaldson speakers do acknowledge any doubts, they simultaneously underline the fact that embryo stem-cell research is 'the only chance'. Indeed, the 'right to hope' and the 'power of hope' (*C4 Evening News*, 16 August 2000) is a central conceptual pillar making resistance to stem-cell research appear morally reprehensible.

In our description of our analytical method we emphasized the importance of paying close attention to any deviation from the main pattern of reporting. Such rigour not only allows researchers to test their assumptions, but can also offer additional insights. The few occasions when journalists adopted a more critical approach to the word 'hope' is a case in point. The word 'hope' is only questioned, in passing, on two occasions in the press coverage within our sample. An editorial in *The Times* mentions 'exaggerated hopes' (17 August 2000) and the *Guardian* reports that: 'She [public health minister] stressed that there was no guarantee of early results, but there was hope – "false hopes" critics predicted' (20 December 2000). The only other example we came across of press reporting which questioned the notion of hope was in an article we excluded from our sample because it was about *adult* stem-cell research, not *embryo* stem-cell research. Noting the way in which hope was treated in this article is nonetheless highly instructive. At this crucial point in time it was vital not only to encourage belief in the possibilities opened up by embryo stem cells but also to deny the potential of adult stem cells, as otherwise the research on embryos might be seen as unnecessary. Anti-hype can, at certain times, be as important as hype in promoting particular scientific agendas. The very title of the article about adult stem-cell research begins by questioning the hope it might offer ('Tantalising dream of a drug-free cure') and final words of the article conclude by declaring 'All these hopes, however, remain just that – hopes' (*Guardian*, 17 August 2000). Just as the term 'only potential' is used to deconstruct the notion of the embryo as a potential human being, so here the term 'just hope' is used to undermine the argument that adult stem-cell research might be a good future source of therapies.

Before concluding our discussion of hope, we should also highlight one other exception to the dominant use of this term – one which appeared in the TV news coverage. A lengthy item on *Channel 4 Evening News* (16 August 2000) began to explore the whole notion of the imperative to hope. In this item an interview with Christopher Reeve (a key promoter of stem-cell research and 'the power of hope') is used in a rather unusual way. The interviewer questions Reeve about a controversial advertisement supporting further stem-cell research which showed Reeve walking (a visual fantasy made possible by digital manipulation of his image). Reeve is challenged about the dangers of creating false hope and is drawn into a brief discussion of the double-edged nature of the concept for many people living with disability. This item illustrates the possibility of a different style of discussion, but one which was marginalized in the bulk of the coverage.

6. *Discussion and Conclusion*

This chapter has analysed how two opposing discourses were projected in the media to assert competing visions of embryo stem-cell research. Our research shows that, as other commentators have noted, there is a great deal more to 'science reporting' than simply communicating scientific facts. The real battle-ground is about the plausibility of diverse visions of utopia and dystopia and about who can claim the authority (in terms of both morality and expertise) to produce a credible version of the future. The chapter has highlighted some of the rhe-torical strategies used in this controversy and examined how the pro-Donaldson perspective sought to characterize itself, and came to be characterized by most journalists, as ultimately more legitimate than the alternative, anti-Donaldson perspective because it was associated with the right sort of people and values.

This chapter contributes to the sociology of expectations and to theories about risk society by offering a case study showing how contrasting representations of the future are promoted in relation to stem-cell research. Brown and Michael argue that there has been a recent shift towards a new 'institutional body language for science', which calls for alternatives to 'expert', detached authority.[13] Our analysis shows how, although the proponents in our study did continue to draw on discourses of authoritative, expert knowledge, this was carefully set within a compassionate framework. The stem-cell example also illustrates the argument by Brown and Michael, that this new science language promotes the rhetoric of transparency and appears to invite 'the public' to take part in the decision-making process. Our analysis allows us to reflect on the extent to which the media provides an avenue for open debate – an issue relevant both to the sociology of expectations and to Beck's appeals for the media to act as 'cultural eyes'. However, we would argue that although the release of the Donaldson report was apparently engineered to facilitate debate prior to the parliamentary vote, the debate was inadequate. Journalists were inconsistent in their willingness to deconstruct and question key terms and presented the debate as a strict binary opposition with little room for ambivalence or 'cautious optimism'. In addition there were key themes missing

---

13   N. Brown and M. Michael, 'From Authority to Authenticity: The Changing Governance of Biotechnology', *Health, Risk and Society* 4 (2002), pp. 259–72.

from the debate. These include the absence of feminist critiques (e.g. concern about women as the source of eggs and/or embryos) and the lack of debate about potential health risks or reflection on the present therapeutic gap.

In part the media were simply mirroring the nature of the debate as it played out in the parliamentary and lobbying process. However, existing studies of media production processes, combined with reflection on what we could observe from the content analysis, also give some clues about the media's limitations as an avenue for this sort of risk debate. The oppositional way in which the news media frame issues as two-sided controversies does not help to break through binary oppositions and may exclude people who fail to offer up 'black or white' positions. A lack of questioning of assertions from scientists may be encouraged by the fact that health and science journalists are often predisposed to accept an optimistic scientific agenda [14] and the financial interest of scientists may be seen as non-news by journalists, because these interests are so commonplace. In particular we wish to highlight the way in which news formats and production cycles mean there is often a lack of time and space to explore more nuanced debates. Journalists' familiarity (or lack of familiarity) with diverse perspectives also influences their lines of questioning. The exceptional example of an interview with Christopher Reeve exploring the problems of 'hope' gives telling insight into the conditions under which more original reporting might be produced. This exchange occurred within a much lengthier format than is usual within news bulletins. It is note-worthy that the interviewer in this case was also a wheelchair user.

In conclusion, we 'hope' that our analysis has contributed to understanding how debates about the future are played out in the pubic domain. The issue of how the future is framed is, perhaps, one of the most important areas of contemporary debate within science and technology studies (STS), sociology of health and illness and media studies. Although news reporting is often tied to news events and what journalists see as 'topical', this does not mean that it has nothing to say about the future. Although the bulk of coverage of stem-cell research in 2000 is firmly tied to on-the-ground events within the surrounding few days, the true focus is on years, even decades ahead. Close attention to how images of the future are constructed in news reporting is, we would suggest, a vital area of research as we seek to understand the construction of scientific and medical developments and policy.

---

14   M. Nisbet, D. Brossard and A. Kroepsch, 'Framing Science: The Stem Cell Controversy in an Age of Press/Politics', *The Harvard International Journal of Press/Politics* 8, 2003, pp. 36–70.

Chapter 10

NEVER TOO LATE TO LIVE A LITTLE LONGER?
THE QUEST FOR EXTENDED LIFE AND IMMORTALITY
– SOME ETHICAL CONSIDERATIONS

Ulf Görman

The human existential condition is often described as characterized by the ability to plan and foresee the future. This involves the capacity to visualize and foresee individual death. In fiction, philosophy and psychology this is widely acknowledged, and often described as a reason for existential anxiety. For instance, Martin Heidegger focuses on human ambivalence by describing the effort to de-personalize this intriguing situation. The personal death of each individual is an ever-present individual trauma, and it is typically human to try to hide or disarm the awareness of this personal fate by objectifying it and talking of it in the third person instead of the first person.1

Zygmunt Bauman describes not only individual life but human culture as para-doxically characterized by the effort to overcome death, based upon the knowledge of the inevitability of death. Bauman looks at a number of human efforts, such as getting acknowledgement or fame, as different ways of transcending death. I guess he could have mentioned reproduction as well. For Bauman, this anxiety and the efforts to handle it, is one of the most important aspects in understanding the human situation. He regards our awareness of mortality as the ultimate pre-requisite for human culture and creativity. The idea of overcoming death in one way or another is meaningful only because of the relentlessness of death. Bauman goes as far as claiming that human culture would probably not exist if we were unaware of our mortality.2

In line with the analysis of Heidegger and Bauman it may be possible to understand modern biomedical efforts as triggered by the ambition to counteract death. Disease means a risk of death, and there is certainly a strong effort in bio-medicine and healthcare to save lives. However, this is connected to a more general endeavour. A wider understanding of human interest in medicine indicates that we want to maintain health. It is widely acknowledged that health is an important value in human life. This is well supported by many studies, and recent investigations in

---

1 This is a central theme in his *Sein und Zeit* (Halle: Niemeyer, 1927), English translation Martin Heidegger, *Being and Time* (trans. John Macquarrie and Edward Robinson; Oxford: Blackwell, 1978 [1962]). This aspect of Heidegger's understanding is intriguing even for those who are sceptical about the heavy metaphysics that characterize his analysis.

2 Zygmunt Bauman, *Mortality, Immortality and Other Life Strategies* (Cambridge: Polity Press, 1992).

Sweden[3] even indicate that health is for many an important aspect of the meaning of life.[4]

But what is health? According to the classical and influential definition by the World Health Organization (WHO), '[h]ealth is a state of complete physical, mental and social well-being and not merely the absence of disease or infirmity'.[5] This is a far-reaching and ambitious understanding, and there is good reason to question whether health in this sense can be achieved except in rare circumstances. However, it may surely be seen as an ideal to strive for, even if it can seldom or never be completely fulfilled. In this interpretation the definition would function as an 'ideal norm' as some ethicists would phrase it.

An ideal norm may be too far reaching and perhaps even misguided. Lennart Nordenfelt, a Swedish philosopher specializing in health, suggests a slightly different and also more cautious definition. According to Nordenfelt, a person is healthy in relation to the extent that he or she in normal circumstances can realize vital goals. One interesting consequence of this way of understanding the meaning of health is that a person can achieve health not only by the fulfilment of goals, but also by realistically adjusting goals to the current situation.[6]

For as long as we know, people have taken an interest in longevity as well as a healthy life. The desire for a long life is recorded in the book of Genesis. According to the narrative of Genesis, the first humans lived very long lives: Adam lived 930 years, Methuselah 969 years and Noah 950. Many religions, including Christianity, respond to the desire to overcome death by offering a world-view which gives hope for or conviction of a life of another kind after death.

The success of biomedicine in dealing with health problems has created expectations that current or future knowledge may be able to change the length of human lifespan on this side of death. This chapter will give some ethical considerations on a few of these efforts. I will make a distinction between four strategies in current biomedicine to handle the knowledge of ageing and death: to normalize ageing, i.e. to prevent premature death; to optimize ageing, i.e. to maintain good health until death; to retard ageing and death; and to eliminate death. They constitute a range of alternatives, from conservative to radical.

1. *Normalize Ageing*

The traditional endeavour of medicine is to cure, care and comfort. Raising ambitions and resources, as well as growth of knowledge, has made medicine

---

3   In this article I use my home country Sweden as an example, in order to illustrate a situation which is fairly similar all over the Western world.

4   Sten M. Philipson and Nils Uddenberg (eds.), *Hälsa som livsmening* (Stockholm: Natur och kultur, 1989). Lena Löwendahl, *Med kroppen som instrument* (Stockholm: Almqvist & Wiksell International, 2002).

5   Preamble to the Constitution of the World Health Organization, adopted by the International Health Conference, New York, 19–22 June, 1946. See www.who.int/about/definition/en/ (accessed 1 October 2005).

6   Lennart Nordenfelt, *On the Nature of Health: An Action-theoretic Approach* (Dordrecht: Reidel, 2nd edn, 1995).

extremely successful during the last 150 years. It was not until the middle of the nineteenth century that medicine and healthcare could achieve substantial changes in the life and life expectancy of humans. The most important reason for this was the understanding of micro-organisms and their importance for specific diseases. This knowledge led to improved hygiene in homes as well as in hospitals. Hygienic measures, such as disinfection, washing and cleaning, improved preservation of food, and disposal of excrements and soil water through sewage systems, were introduced. As a result of these and many other steps, the rate of deaths in childbirth and epidemics diminished. Another important step was taken in the middle of the twentieth century, when antibiotics were discovered and came into frequent use. Infections could now be effectively controlled. The likelihood of early death was radically diminished.

Average lifetime in Sweden at the middle of the eighteenth century was 34 years for men and 37 for women, while average life expectancy at birth in 2004 was 78.3 years for men and 82.7 years for women. The largest change has been in the death rate at early ages. Death at higher age has been more difficult to master, and when a higher number of people can reach their sixties cardiovascular diseases and cancer become more frequent causes of death. The maximum lifespan as we know it now is around 120 years.[7] This age is exceptional and has been reached only by a very small number of individuals.

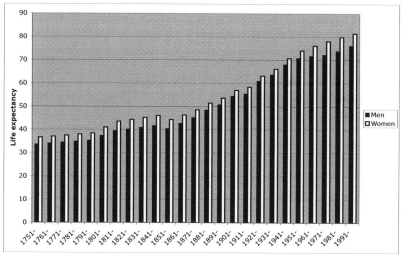

Figure 1: Mean life expectancy at birth in Sweden 1751–2000. Based upon statistics from Statistiska Centralbyrån, Sweden, *Befolkningsstatistik, Återstående medellivslängd för åren 1751–2004*, www.scb.se/templates/tableOrChart____ 25830.asp (accessed 1 October 2005)

Figure 1 shows how the mean length of life has more than doubled over the last 150 years. From the middle of the nineteenth century lifetime started to rise

7   Roy L. Walford, *Maximum Life Span* (New York and London: W.W. Norton, 1983).

substantially. Life expectancy is still increasing, although at a lower rate. This remarkable change is probably the result of a number of other changes in society as well, such as a raised standard of living, an effort in society to diminish risks, etc., but the success of medicine in preventing disease and premature death has made substantial contributions in this respect.

The effort to prevent premature death has been the focus of a number of societal and medical strategies for many years. The likelihood of early death has diminished substantially. There is no doubt that these efforts correspond to needs and demand from the people involved. Premature death has always been seen as an unlucky fate. Few would want to see the return of the situation where an infection was a life-threatening disease for most people. There is also good reason to believe that not many people living in the Western world would like to go back to a situation where a family needed to have five or more children, of which most were expected to die at a young age, in order to have some safeguard against infirmity when growing old.

The idea of a premature death rests on a normative concept of normal ageing and death. This idea can certainly be contested. What is considered as normal has certainly changed over the years and will probably continue to change. But in general the effort to prevent premature death corresponds well with human values and the expectations of almost every human being. The results of these efforts have been tremendous. They should be welcomed from an ethical point of view, and so should the continuous struggle for this goal.

2. *Optimize Ageing*

Now early death has been mastered, the focus in medical practice, research and development has successively moved to new health problems. The most important causes of death in Sweden in 2000 were cardiovascular diseases (46 per cent) and cancer (24 per cent), which means that these two groups of diseases cause 70 per cent of all cases of death. Cardiovascular diseases show a downward trend. The number of deaths caused by cardiovascular diseases decreased by as much as 40 per cent from 1987 to 2000.[8] A large number of steps have contributed to this development, such as medication and surgery, as well as physical exercise and dietary advice.

One visible change is the fact that ageing is no longer considered a natural and inevitable process, but great efforts are devoted to the idea of maintaining health during the whole lifespan. This goal will partly be achieved by 'manufactured time', i.e. additional lifetime created by different kinds of technical steps.[9] Infirmities that were earlier perceived as natural and inevitable results of ageing

8   From 1987 to 2000, the death rate in cardiovascular diseases for men decreased from 360 to 207 per 100,000 men at the age of 15–74. For women the corresponding death rate decreased from 133 to 88. *Dödsorsaker 2000. Socialstyrelsen: Sveriges officiella statistik, Hälsa och sjukdomar 2002:4*, www.sos.se/FULLTEXT/42/2002-42-4/sammanfattning.htm (accessed 1 October 2005).

9   S. Jay Olshansky and Bruce A. Carnes, *The Quest for Immortality: Science at the Frontiers of Aging* (New York and London: Norton, 2001).

are now in many cases understood as specific diseases that may be treated and perhaps even cured. An increasing number of diseases can be retarded by medication. Many medical treatments, including surgery, are refined and applied at higher ages. Spare parts to restore bodily functions are used more frequently and at higher ages, as well as organ donations. The recent fight against dementia is another step in the same direction. Big hopes are oriented towards stem cells as a new tool for the future.

This change of focus has been challenged, among others, by a number of scholars, including the influential American ethicist Daniel Callahan, who has a particular interest in problems of death and extension of life. The claim Callahan makes is that in developed countries we ought not to try to extend our lives beyond the current state. 'Neither the human species as a whole, nor most individuals, need more than the present average life expectancy in the developed countries (the mid-seventies to low eighties) for a perfectly satisfactory life. This idea of a steady-state life expectancy at its present level would establish happily, a finite and attainable goal: "Enough, already." '[10]

Callahan clearly wants us to make efforts to prevent and diminish illness and untimely death at an early age. But: 'Once people have made it past the point of a premature death (a point I would set in the range of sixty-five to seventy years), then the highest priority for medicine should become not to avert death but to enable people to live as comfortable and secure a life as possible.'[11] Because of this, technological medicine will be most justifiably used for what he calls adults, i.e. people between 21 and 65, and he recommends limited use of high-tech medicine and expensive healthcare for those who are in their late seventies or older.[12] He terms this view 'life cycle traditionalism', and he explains that '[i]t is based on the biological rhythm of the life cycle as a way of providing a biological boundary to medical aspiration'.[13] Consequently, in Callahan's view, to try to extend individual life beyond this rhythm would be a distortion of an important stage of life. It would mean to take away from elderly people a part of their life experience. Instead all adults should learn to understand the meaning of this final part of life. This is also what biomedicine should focus on.[14]

The claim of Callahan raises a number of questions. One of the claims Callahan makes is that aging and death is a natural part of human life. 'Death is an inescapable reality of human life and always will be ... Our humanity is, in great part, defined by our willingness to accept and live with death.'[15] Callahan claims that we should accept death and, even further, that this acceptance constitutes what

10   Daniel Callahan, *False Hopes: Why America's Quest for Perfect Health is a Recipe for Failure* (New York: Simon & Schuster, 1998), p. 133.

11   Callahan, *False Hopes*, p. 256.

12   Callahan, *False Hopes*, p. 255.

13   Daniel Callahan, 'Aging and the Life Cycle: A Moral Norm?', in D. Callahan, R.H.J. ter Meulen and E. Topinkova (eds.), *A World Growing Old: The Coming Health Care Challenges* (Washington, DC: Georgetown University Press, 1995), pp. 20–27.

14   Daniel Callahan, 'Death and the Research Imperative', *New England Journal of Medicine* 34 (2000), pp. 654–56.

15   Callahan, *False Hopes*, p. 21.

it is to be human. In one sense this is incontestable: human mortality is 100 per cent, and there is no single known exception to this fate. The inevitability of death is also an integral part of the human existential condition. I agree that it is an important aspect of human life not only to fight death, but also to come to terms with it. I will come back to this point. However, when we talk about a stage of life being natural, this can be understood as the claim that there is a genetic base for this life stage. In this way human growth from embryo over foetus, childhood and adolescence to a fertile adult human being is a natural process, controlled by different genes that express themselves at certain stages of development and use hormones to trigger different bodily processes such as differentiation and growth, as well as the finishing of such processes.

Now, is ageing and death a stage of life in a similar sense? Are there ageing genes that express themselves when an adult has reached a certain stage of life, and create an ageing process, the maturity of which is the death of the organism? This seems not to be the case. According to the dominating theories of ageing, this is not the result of an active, genetic or other, ageing process. Instead, ageing is a result of a number of different factors, such as deficiency in bodily functions, a gradual decrease in the bodily self-repair system, injuries and similar things. Gradually the organism becomes worn out. Olshansky and Carnes sum up the prevailing current understanding by stating: 'the biological clocks that are present within an organism exist for one reason only – to support life, not destroy it'.[16] The time for different functions of the organism to be worn out is extremely divergent among individuals. Some persons in their eighties may be physically younger than others in their fifties.

Olshansky and Carnes explain the situation by introducing the concept of 'warranty period', which is the period needed for successful reproduction, including time for childcare and perhaps even grand-parenting. '[S]enescence-related diseases and disorders observed in the postreproductive period are unintended by-products of selection acting upon genes participating in biological processes important earlier in the life span ... This view of senescence also implies that aging is not an unnatural disease, but is instead a natural by-product of survival extended into the postreproductive period of the life span.'[17] The huge difference between individuals in terms of their longevity is probably connected to genetic variation as well as differences in lifestyle. But the idea that there should be a natural biological clock for ageing and death in our organisms that we ought to adjust to, does not seem to fit with current medical knowledge.

A claim that Callahan clearly makes is that 75–80 years is an optimal length of life. But this seems to be arbitrary. 150 years ago a person could feel privileged if he or she had lived to 50 and still enjoyed a good life. Today many are vital at the age of 80 or more, and many of these enjoy and benefit from manufactured

16   Olshansky and Carnes, *The Quest for Immortality*, p. 69.

17   S. Jay Olshansky, and Bruce A. Carnes, 'In Search of the Holy Grail of Senescence', in Stephen G. Post and Robert H. Binstock (eds.), *The Fountain of Youth: Cultural, Scientific, and Ethical Perspectives on a Biomedical Goal* (New York: Oxford University Press, 2004), pp. 133–59. Quotations from pp. 140 and 141.

time through medical interventions. Mean lifetime expectancy is still increasing. However, the quality of life and the length of life is strongly divergent. To define 75–80 as the optimal lifetime is arbitrary.

Another claim that Callahan clearly makes is this: He says that old people have had enough experiences, and that it is fair or just to give them low priority compared to those who are younger. In response to this I will not go into the question of just distribution of welfare and health resources.[18] The main problem seems to be whether we should strive at the development and use of knowledge and techniques for healthcare at higher ages or not.

When it comes to problems of health, each individual should be protected in relation to needs. This principle has rightly been used to give special support for handicapped people, as well as in healthcare in general. How can this principle be used in relation to the development of treatment for elderly persons? In order to deal with that question we may look at the idea of a good life for an ageing person.

British gerontologist Tom Kirkwood understands good ageing to be the harmonious decline of all organ systems, resulting in a growing probability of death.[19] According to this definition, an optimal ageing process is not to keep bodily functions undiminished until the end of life, but instead a balanced weakening of the human body. In his studies, Kirkwood is interested in the mechanisms of cellular ageing, and his definition does not say anything about the mental relation to ageing and death. In this respect, I want to refer to a definition from Canadian ethicist Wayne Sumner. He regards the two concepts of happiness and life satisfaction as identical and describes life satisfaction as a positive cognitive and affective response on the part of a subject to the conditions or circumstances of her life.[20] According to his definition, life satisfaction does not depend directly on the external fulfilment of certain desires or preferences, but consists instead in the mental acceptance of how things are. Sumner's understanding of life satisfaction corresponds well with Nordenfelt's definition of health as dependent on the extent to which a person in normal circumstances can realize vital goals. According to Nordenfelt, a person's goals are not simply given, but expectations may be adjusted to the situation of the individual.[21]

This way of looking at health, good life and ageing indicates that optimization of ageing is not necessarily dependent on full physical capacity, but has a strong mental component, involving the cognitive and affective acceptance of a certain condition of life. This explains why some persons may feel healthy, even if they to some extent are infirm.

Now, age-related diseases such as stroke, cancer, Alzheimer's, Parkinson's, diabetes, and arthritis become more frequent as more individuals live until they become 65 or older. Diseases such as these will in many cases prevent harmonious

18   For discussion on this question see, among others, Norman Daniels, *Just Health Care* (New York: Cambridge University Press, 1985), and Tom L. Beauchamp and James F. Childress, *Principles of Biomedical Ethics* (Oxford: Oxford University Press, 5th edn, 2001), chapter 6.

19   T.B.L. Kirkwood, *Time of Our Lives: The Science of Human Ageing* (Oxford: Oxford University Press, 1999).

20   L.Wayne Sumner, *Welfare, Happiness, and Ethics* (Oxford: Clarendon Press, 1996).

21   Nordenfelt, *On the Nature of Health.*

ageing, as they diminish quality of life in a way that the individual is not prepared for and finds difficult to accept. All too often these diseases, just like many others, are detrimental to the person affected, as well as to friends and relatives. They are not different from the conditions that we have already been able to master.

So, harmonious ageing may be a balance between acceptance of the weaknesses we cannot do anything about and at the same time medical assistance of the kind that helps the ageing person to combat such weaknesses that can be mastered. From the aspect of the individual this means that the individual has reason to ask for assistance in all circumstances that prevent the realization of vital goals in life. At the same time the individual has reason to adjust these goals in a realistic way to his or her capacities and situation in life. Society has reason to give special support to those who are weak, handicapped by diseases, including those diseases that come by age. This includes the effort to invest resources in developing new knowledge about those diseases that we are unable to handle but may find a solution to by further research. Healthcare has reason to give support at all stages of life when quality of life can be raised. We should not define a normal length of a lifespan and use that to treat diseases, but rather look at the possibilities for life satisfaction. We should welcome current and future knowledge that may remove, postpone or mitigate diseases that prevent life satisfaction.

3. *Retard Ageing and Postpone Death*

More than 20 years ago geneticist Michael R. Rose managed to show that it was possible to extend, even double, the lifespan of fruit flies, Drosophila, from about 60 days to 120, by selective breeding through a number of generations.[22] The success of Michael Rose has created great hopes for longevity, but what has often not been noted is the fact that the achievement of Rose was made in artificial surroundings. In real life, predators threaten Drosophila, and the quick breeders are those who are able to reproduce. Rose created artificial surroundings, where those that lived longest and were slow breeders were artificially selected (by Rose) for reproduction.

Anyway, the achievement of Rose and others have initiated partly successful experiments on different animals aimed at extending lifespan by direct genetic modification, and even created hopes that this may be possible for human beings in the future. Many have found it difficult to apply experiences from Drosophila to humans, but others have been more optimistic. One example of this endeavour is Aubrey de Grey, who foresees that by genetic engineering of humans he will be able to postpone death and in the future expects life expectancies to reach 1000 years.[23]

De Grey focuses on a number of reasons for age-related physiological decline,

---

22   Michael R. Rose, 'Laboratory Evolution of Postponed Senescence in Drosophila Melanogaster', *Evolution* 38 (1984), pp. 1004–10; Michael R. Rose, Hardip B. Passananti and Margarida Matos (eds.), *Methuselah Flies: A Case Study in the Evolution of Aging* (New Jersey: World Scientific, 2004).

23   Aubrey D.N.J. de Grey, 'An Engineer's Approach to Developing Real Anti-Aging Medicine', in Post and Binstock (eds.), *The Fountain of Youth*, pp. 249–67.

and on finding measures to repair this decline rather than to stop the damage it causes. He identifies a small number of major causes for decline, such as cell loss, mutations, cell senescence, enzyme deficiency through genetic defects, and immune system decline. The policies for repair that he suggests go beyond current medicine and include several versions of gene therapy, in order to give cells extra enzymes that can prevent enzyme deficiency, and above all to try to stop age-related tumours.[24]

The vision of de Grey illustrates the possibilities, but also the problems connected to the radical extension of human lifespan. Some of the presuppositions for human life need to be changed substantially in order to achieve this goal. The goal he is striving for may be the most interesting question. Let us come back to that in a moment. But there are a few evident problems that have to be faced. One of these is whether we know enough about the interplay between different genes and the possible effects and side effects of genetic changes. This is connected to the balance between benefit and risk. All interventions in the human body are connected to some kind of risk. In general these risks can be expected to be larger when the intervention is more far reaching. If the risks are well known they may be balanced against the benefits, but if they are not well known enough, it is not even possible to evaluate the balance.

History is full of examples where interventions have had unexpected side effects. The heavy use of oestrogen for women around and beyond menopause is just one of the cases where unforeseen problems have emerged. Use of oestrogen has proved to increase the risk of heart disease, stroke and blood clots, as well as breast cancer.[25] When it comes to the vision and proposed interventions of de Grey, we can expect to have very limited knowledge of such side effects. We have only limited knowledge about the complex interplay between the expressions of different genes. Perhaps a genetic change in order to radically extend human lifespan will as a side effect have serious diseases.

Another question is how large are the risks that are worth taking. In general it is appropriate to relate the risk involved in an intervention to the character of the problem that can be solved. The hope of curing a life-threatening illness may be a reason to take large risks. We accept severe treatment against a deadly cancer, while we rightly think that a less serious condition needs a more restrictive risk acceptance. Likewise, most people agree that it is a good thing to take risks in order to cure a disease or a handicap, while enhancement is much more questionable.

My contention is that whenever such unknown risks appear, we should be very careful if the aim is to enhance rather than to cure. In the case of a disease, we know the suffering involved. People are willing to take the risk of certain side

---

24   See de Grey, 'An Engineer's Approach' and the references to other works of de Grey made there.

25   J.E. Rossouw, G.L. Anderson, R.L. Prentice, *et al.*, 'Risks and Benefits of Estrogen Plus Progestin in Healthy Postmenopausal Women: Principal Results from the Women's Health Initiative Randomized Controlled Trial', *Journal of the American Medical Association* 288.3 (2002), pp. 321–33.

effects, depending on the severity of their situation. Ethical standards for clinical practice take this into account. But when it comes to enhancement, such as the achievement of an extraordinary length of life, we may know very little about the side effects or how this extraordinary situation will affect the quality of life of the person involved. Besides, this will cause justice problems if only a few persons can achieve a few hundred years of life.

As human beings we are basically relational. We grow up in relations, we live in relations, we even understand ourselves in terms of relations, and if we try to be independent we still cannot be but involved in relations of many kinds. The relational character of human life is part of the idea of a good life and the idea of quality of life. Modern individualism tends to overlook the relational character of most aspects that form us as human beings. The efforts to radically postpone death and retard ageing also seem to be radically individualistic visions. This involves the risk that by achieving these far-reaching goals, the long-living beings also miss an important part of what it is to be human. If, on the other hand, many or most humans will live a few hundred years this will cause great demographic problems. In its turn this may affect human reproduction in ways we can only imagine.

The main question, however, with the idea of a radical life extension, seems to be of another kind. The human existential condition, as I described it at the beginning of this chapter, involves the limitation of life and the inevitability of death as well as the longing to overcome death. This is not only or mainly a bad situation. Bauman may be right that this paradox is the prerequisite for human culture and creativity. What will happen if we manage to change this character of human existence, for some or for all? We don't know. A number of aspects that may be central to a good life are called into question if human life is 'enhanced' by radically changing the human condition. If a good life involves a harmony between our ambitions and what we can achieve, if a good life involves fruitful and mutual relations to others, if a good life involves pleasure as well as longing and desire, development through mistakes and suffering, how will these aspects of a good life be fulfilled if the human existential condition is radically changed?

Still another option offered in order to extend individual life is cryonics. This is the label for hopes oriented at freezing the whole body, or sometimes only the head, before death, or in an intermediate state where the legal definition of death is met but not yet the biological.[26] My preliminary contention is that this may not easily be disregarded as innocent optimism. Instead, people are encouraged to pay some \$28,000–120,000 to have their bodies frozen.[27] There is no evidence that they can be successfully brought back to life, except the vague hope that future scientific knowledge will make this possible, and that their dead or dying bodies may stay preserved during freezing and future thawing. At this moment this seems to be a kind of fraud. In my view it is one of society's tasks to protect its citizens against fraud, and because of this it should not be allowed by society.

---

26   Robert C.W. Ettinger, *The Prospect of Immortality*, www.cryonics.org/book1.html (accessed 19 June 2005).

27   Cryonics Institute, *Comparing Procedures and Policies*, www.cryonics.org/comparisons. html (accessed 19 June 2005).

Humans have longed for long life and immortality since the beginning of recorded history. This is richly reflected in fiction and poetry. However, many, perhaps most, of the fictional stories about radically extended life or immortality have been dystopias. One of the most well-known examples is the tale of Jonathan Swift about Gulliver's visit to Luggnagg, where sometimes a child was born with a circular spot on the left side of the forehead, an infallible mark that it should never die. Initially struck with delight about the idea of immortality, Gulliver soon discovers that the lives of these so-called struldbrugs, far from being blissful, are very unfortunate. From the age of 30 they become dejected at the dreadful prospect of never dying. From about 80, when they are older than their fellow citizens, they become incapable of friendship and dead of all natural affection. From about 90 they lose their teeth and hair and distinction of taste. They are plagued by diseases and become senile. Having learnt about the fate of the struldbrugs, Gulliver loses his appetite for perpetuity of life.[28]

Another well-known narrative about endless human life on earth is the story about Ahasverus, the wandering Jew, who is doomed to restless wandering around the world until the day of judgement, without being able to die. The story has inspired many authors, for instance Goethe, Wordsworth, Kipling and Lagerkvist. In such stories an endless life is depicted as a curse, characterized by isolation, uneasiness and meaninglessness. In these and many other fictions and poems, the prospect of extended life and immortality has often either been connected to fear for loss of mental capacities, or to a meaningless and lonesome existence.[29]

4. *Eliminate or Overcome Ageing and Death*

'Transhumanism' and 'posthumanism' have become labels for efforts to take even one more step, to eliminate ageing and death completely, at least for certain individuals.[30] How this might be achieved is often very vaguely described. But it is also possible to find some elaborated efforts in this direction.

In his book *Becoming Immortal*, embryologist Stanley Shostak considers the possibility not only of extending life expectancy, but also of creating immortal humans.[31] The possibility of extending the lifespan of a number of organisms is not too difficult a task, but the prospects may be more limited in humans who are already the longest-living mammals. However, to create biological immortality would require more, and Shostak presents a solution. Can death be overcome by biotechnology? Yes, Shostak says, this can be achieved through a combination of cloning and stem-cell technology. According to his proposal, this requires a com-

---

28 Jonathan Swift, *Gulliver's Travels, into several remote nations of the world* (London, 1726).

29 Many more examples of literature on longevity can be found in Carol C. Donley, 'Primary Literary Sources on Prolongevity', in Post and Binstock (eds.), *The Fountain of Youth*, pp. 433–43. See also Leon Kass (ed.), *Being Human: Core Readings in the Humanities* (New York: W.W. Norton, 2004).

30 See, for instance, www.transhumanism.org (accessed 1 October 2005).

31 Stanley Shostak, *Becoming Immortal: Combining Cloning and Stem-Cell Therapy* (Albany, NY: SUNY Press, 2002).

bination of two genetic modifications. One of these is to introduce an improved stem-cell generator, an endless source of stem cells that regenerates bodily functions when something goes wrong in the copying process. The other modification is to suppress sexual maturity. By this, human development can be stopped in the pre-pubescent stage of development, at the physiological age of about 11. Immortals will be forever young, sexually immature and sterile. The vision of Shostak will not be able to make us immortal, but this can be secured for our cloned children.

A number of arguments can be raised against this vision. One is of course that it comes close to the dystopias discussed above. This problem has already been discussed. In addition to that, the question is whether these constructions will be human at all. I would like to argue that these everlasting immortals are not humans in the full sense of our understanding. They will not only lack the knowledge of a future but unpredictable end to life. They will also lack the possibility of sexual attraction, sexual relation, and reproduction. Unless their intellect has been adjusted as well, they can be expected to understand this limitation in their life. Either they will understand that they lack an important existential dimension of life, or they will not even understand this limitation. In both cases they cannot be considered humans in the full sense. In any case I want to argue that it is a violation against these beings to produce them on purpose.

The vision of Shostak is an extended version of reproductive cloning. Reproductive cloning is the idea of using assisted fertilization in combination with nuclear transfer in order to create an individual in the next generation, genetically identical with an already existing being. The main argument for rejecting this possibility is that it is a violation against the cloned individual. The outrage consists in the fact that he or she is produced to be a copy of another individual, born to fulfil a personal dream of a life in the next generation. This takes away from them the basic right to freedom and dignity. Existentially and psychologically this violation may result in a burden of life.

5. *Conclusions*

I suggest that improved treatment of disease and handicap at all ages should be welcomed. As a secondary effect, this may in many cases result in an extension of individual lives. However, the direct aim at life extension through modification of certain basic human properties, such as genetic enhancement, should be looked at with suspicion. It may involve exaggerated risks, but the main argument is that it may diminish the existential character of what it is to be human, involving relation to others as a central aspect. To intentionally produce humans so that they become immortal is a violation against these individuals because they lack an important existential dimension of human life.

Chapter 11

GENETIC PERFECTION, OR FULFILMENT OF CREATION IN CHRIST?

Maureen Junker-Kenny

'Be perfect, as your heavenly father is perfect.' (Mt. 5.48)

Surely, if ever there was a perfection movement in the history of human self-understandings, it has to be Christianity. The spirituality it brought forth was, many would say, relentlessly geared towards perfection. Only recently, and helped by cultural critics such as Nietzsche, have we thrown off the life-denying drive towards spiritual perfection that was engrained in the Christian mentality at the expense of human flourishing and self-realization in this life.

So why is it that suddenly Christian ethics is turning against 'perfection', once this radically Christian idea is coming towards us from other quarters, notably from the bio-sciences applied to perfecting human nature not just at the surface but at the roots? What can be wrong about striving to perfect the physical and possibly mental bases of our individual selfhood? Better memories, fitter bodies, multi-purpose enhanced abilities, possibly even greater religious fervour, outreach, sensibility for the divine? Why should it be necessary to construct an alternative between such steps towards perfecting human nature, and the Christian belief in Jesus Christ as the ultimate fulfilment of what God intended in creating human beings as partners for God's love?

I shall explain in the following steps my contention that the thrust towards genetic self-perfection is morally, anthropologically and theologically mistaken. If it is true that one can detect a new drive towards perfection in quite different symptoms,[1] how is practical philosophy responding to it? What can be said about self-perfection from a secular ethical viewpoint (section 1)? In view of Christianity's history of thinking and practising its version of perfection, what was its response when the idea of human self-perfection emerged in the Enlightenment as a serious contender for the Christian ideal of discipleship and Christ-like living (section 2)?

---

1 Films like *Gattica* at the level of popular culture are matched by debates in expert cultures on the desirability of a new age of 'anthropotechniques' (P. Sloterdijk) where Nietzsche's dream of the *Übermensch* finds its biological realization. No longer ends in themselves, present-day humans are just the bridge to this new stage of life. Beyond science fiction and a philosophy of the human person as her own experiment, however, it is at the institutional level of insurance calculations and government policies that perfection is more than a lofty thought. The medicalization and seeping quality-control regime of pregnancy and the restriction of the right to privacy in allowing insurance companies to ask for genetic test results clearly indicate the direction of future interests.

1. *Contemporary Philosophical Critiques of Genetic Enhancement*

One can always find supporters among philosophers and even feminists for altering human, i.e. male and female, nature from what they identify as their particular drawbacks. The 'equal opportunity' argumentation of Alan Buchanan and his co-authors in the Rawls line of justice[2] offers an example for justifying such intervention of parents into their children's genes. The cause of fighting for social justice and equality is moved from the civic arena of changing social structures (or re-evaluating orders of recognition, e.g., of motherhood or parenthood) to the interior scene of genetic aptitude. Yet, other philosophical responses exist which theology should not ignore. I shall examine two of them.

a. *The Moral Imperative of Symmetry*

The first critique, by Jürgen Habermas, is based on moral reasons. For him, the symmetry aspired to in the parent–child relationship is breached by unilateral irreversible domination when parents, presuming to know their child's best interests, wish to enhance his or her genes. Habermas deepens Kant's morality of respecting the equally original freedom of the other with Kierkegaard's concept of being able to be oneself (*Selbstseinkönnen*).

> [As] the designer makes himself the *co-author of the life of another*, he intrudes – from the interior, one could say – into the other's consciousness of her own autonomy. The programmed person, being no longer certain about the contingency of the natural roots of her life history, may feel the lack of a mental precondition for coping with the moral expectation to take, even if only in retrospect, the *sole* responsibility for her own life.[3]

Habermas's critique highlights a hitherto unquestioned premise: Personal identity in the sense of being able to take responsibility for one's own life needs to be developed from a 'given', 'contingent' natural basis as opposed to one designed. The difference between educational determinations through socialization and genetic ones is that a grown-up child can choose to distance herself from her upbringing while genetic interventions confront her with permanent third-party choices which concern the core of her self-understanding. Confronted with the future intention of parents to determine the genetic specificity (*Sosein*) of their child, Habermas sides with fallibility and finitude versus feasibility, with autonomy as responsibility rather than choice, and with the self-determination of the other as a limit to everyone's exercise of power. The breach of unconditional acceptance of the selfhood of their children is Habermas's main point of objection in the case of enhancement.

The question of therapeutic genetic intervention falls under the different criterion of preventing a life-threatening illness and is not dismissed as under-

---

2 A. Buchanan, D. Brock, N. Daniels and D. Wikler, *From Chance to Choice: Genetics and Justice* (Cambridge: Cambridge University Press, 2000).

3 J. Habermas, *The Future of Human Nature* (Cambridge: Polity Press, 2003), pp. 81–82, original emphasis; finitude, pp. 89–90. I have analysed his argumentation in greater detail in 'Genetic Enhancement as Care or as Domination? The Ethics of Asymmetrical Relationships in the Upbringing of Children', *Journal of Philosophy of Education* 39 (2005), pp. 1–12.

mining the child's 'possibility of being herself'. But the ethical problems of the experiments necessary for this route as well as the much-debated and unresolved questions about definitions of illness (e.g., genetic predispositions, late-onset disorders) and about the operative criteria (biological disfunction, physicians' consensus, the patient's subjective view on their suffering or lack of suffering, etc.) are not touched upon. In his recent publication, *Es wird ein Mensch gemacht. Möglichkeiten und Grenzen der Gentechnik* (Berlin: Rowohlt, 2003), the scientist Jens Reich concludes each of his chapters explaining the genome, stem cells, cloning, pre-implantation genetic diagnosis, prenatal diagnosis and germ-line engineering with a fictional dialogue. Discussing the hope of abolishing severe hereditary illnesses once and for all through germ-line therapy, he says the problem is where to draw the line between such inherited debilitating illnesses and mere anomalies or harmless genetic variants: 'Take excess weight... And from what degree onwards is being short-sighted a severe defect rather than a harmless anomaly?' (p. 170). In view of the difficulty of drawing secure limitations between medical and cosmetic indications, and of the fact that there have been parents for whom, e.g., the diagnosis of a cleft palate has been the reason for aborting their future child, the separation between avoiding illness for the child's sake, and only accepting a perfect child is less clear-cut than it seems. Reich's dialogue reiterates the anticipation of symmetry in the relationship of parents and children: 'If one concedes to parents the right to have perfect children, would one not have to equally concede to children the right to have perfect parents? Just imagine if children could get rid of their parents just because for them they are too fat, not sporty enough, too dumb or short-sighted. This sounds perverse, but it would not be very different if parents could discard their embryos for exactly these reasons' (p. 170). The fiduciary relationship that the law assumes between parents and children lives off the unquestioned premise that they can be trusted to deliver more than the legal minimum of duties for their offspring. Parental perfectionism and its consequences, however, such as the development of guidelines that specify which defects cannot be grounds for an abortion, show that this assumption may not always be justified. This is all the more reason for concern since a parental attitude that puts conditions on accepting their child even undercuts the relationship among citizens as mere strangers who are always accorded their human dignity, sometimes despite their performance.

There is a double reduction in focusing on defective genes: the (future) person to her body, and her body to her genes. This is incompatible with the principle of human dignity which holds irrespective of bodily constitution, performance, and even moral integrity. It is an asymmetric principle in that it recognizes also those who do not recognize the human dignity of others, including, e.g., war criminals. But the readiness to uphold this principle of distinguishing between a person's acts and their innate dignity even when it is not reciprocated and when it has been violated by themselves seems to give way to other concerns when it comes to parental hopes for excellence. Even when they are reinterpreted as wishes for their children's happiness, the lack of distinction between the empirical features and the person of the child remains problematic. While in the case of genetically inherited illnesses the scientists' motivation to help and the parents' wish to spare their child unnecessary suffering are morally evident, one cannot ignore that the debate takes place on the background of an increasingly economy-led view of whatever public healthcare system exists. Instead of building a society that reckons with and provides for the frailty endemic in the different stages of the human condition, the risks and interests of individuals are set up against each other and the costs pushed back to the individual unit of the family. While Habermas does not pursue the moral questions raised by genetic intervention for reasons of therapy, not enhancement, he would endorse the principle of symmetry on which Jens Reich bases his objections.

b. *Objections from Philosophical Anthropology*

From a different, but complementary, angle Ludwig Siep expresses doubts about the wisdom of turning to genetics to assist us in fulfilling the goals of our lives. The relationship between genetic make-up and life-project is more complicated than admitted by perfectionists. Instead of defining ourselves through optimal genes, our real potential lies in the human capacity for learning and in finding fulfilment by succeeding in spite of obstacles, including our genetic dispositions.

> A considerable part of the experience of what is valuable, satisfying and fulfilling results from the mastering of problems and the fulfillment of tasks recognized as necessary and meaningful... It is less in the realisation of projects than in the discovery of what we are good at, and in the often surprising insights into what we really care about that we encounter meaning in life. And these discoveries are by no means always along the line of our dispositions.[4]

If these anthropological insights are true, help is not to be found in genetic enhancement (which will always remain incremental and finite), but in knowing oneself and one's limits:

> It is neither realistic nor advisable to regard our body and our environment as mere material for self-projections. If this thesis is true, then the long-term goal to give as much information and possibilities for 'optimising' one's genetic heritage must be viewed skeptically... How many have produced major achievements without the support of specific dispositions or even against them? How much fulfillment grows from activities for which we were not particularly endowed as long as we were able to know our limits?[5]

The courage to take a stance on one's individual limits is more likely to occur in a culture of thinking that accepts the fragility and vulnerability of human life. The Jewish philosopher Hans Jonas locates human dignity precisely in this frailty in opposition, e.g., to the utopian sympathies of Ernst Bloch's philosophical transfor-

---

4   L. Siep, 'Genomanalyse, menschliches Selbstverständnis und Ethik', in L. Honnefelder and P. Propping (eds.), *Was wissen wir, wenn wir das menschliche Genom kennen?* (Köln: Dumont, 2001), pp. 196–205 (199).

5   Siep, 'Genomanalyse', pp. 199–200. A similar intuition can be found in Isaiah Berlin's judgment that 'we are free when we know our limits'. From a variety of philosophical perspectives Siep's conclusion can thus be endorsed: 'With regard to goals for the future, it should be the subject of public discussion how good it is for the human person to measure her natural "start conditions" solely by the scope of possible self-projects, and to attempt to optimize them' (p. 203).

mation of the Jewish messianic heritage.[6] It is worth investigating where modern Western thinking at the turning-point of the Enlightenment into the Romantic Age located the essential conflicts within the human self and their chances of resolution.

2. *Enlightenment Debates on 'Perfectibility'*

Can theology offer any further insights beyond the convergence of these moral and anthropological considerations? By going back to the first round of the 'perfectibility' debate in and after the Enlightenment, we shall encounter analyses of human subjectivity that show just how shallow some of the currently held understandings of selfhood are. The theological backdrop to Johann Gottfried Herder's view on perfectibility is the idea of humans as being made in the image of God. Later, Schleiermacher develops his limit reflections on human perfection in the context of his Christology.

a. *Diversity as the Mark of Humanity, the Antithetical Structure of Subjectivity, and 'Romantic Expressivism' (J.G. Herder)*
At first glance, Herder seems to be an advocate for the need to strive from rudimentary natural endowments towards perfection. His observation that at infancy the human being is 'the most orphaned child of nature' concludes in the call that 'we are not yet men, but are daily *becoming* so'.[7] Is there anything in this view to separate it from Nietzsche's contempt for the 'last man' beyond whom we are to progress to a new dawn? Is 'not yet being, but *becoming* human' not the clarion-call of the progress mentality of Western modern thinking well before it set out to achieve the technical means of creating from the present type of humankind a future, fail-free version?

i. *Perfection as specifically human diversity*
Surprisingly, realizing the poverty in instincts of humans compared with their animal co-creatures, Herder does not turn his gaze upwards towards a future race that has compensated for nature's initial miserliness; he directs it sideways to the potential that the presence of fellow-humans signifies. Faced with our natural defects, his outreach is horizontal, not vertical. 'Naked and bare, weak and in need, shy and unarmed',[8] the orphaned state of the child's senses and skills is counterbalanced by those who help us to achieve our 'spiritual birth', connecting us with our 'parents, teachers, friends,... countrymen and their forefathers; and lastly with the whole chain of the human race'.[9]

---

6   Cf. C. Wiese, '"Dass man zusammen Philosoph und Jude ist..."', Zur Dimension des Jüdischen in Hans Jonas' philosophischer Ethik der Bewahrung der "Schöpfung"' in J. Valentin and S. Wendel (eds.), *Jüdische Traditionen in der Philosophie des 20. Jahrhunderts* (Darmstadt: Wiss. Buchgesellschaft, 2000), pp. 131–47.

7   Herder's *Essay on the Origin of Language*, in J.J. Rousseau and J.G. Herder, *On the Origin of Language*, trans. J.H. Moran and A. Gode (New York: Bergman, 1967), pp. 107–108, is quoted from W. Pannenberg, *Anthropology in Theological Perspective*, trans. M. O'Connell (Philadelphia: Westminster, 1985), pp. 43–60 (44), original italics.

8   Herder, *Origin*, pp. 107–108, in Pannenberg, *Anthropology*, p. 43.

9   Herder, *Outlines of a Philosophy of the History of Man*, trans. T. Churchill (London, 1800; Chicago: University of Chicago Press, 1966), IX, pp. 227–29, quoted in Pannenberg, *Anthropology*, p. 46.

This educational dependency on 'tradition and learning' is something to be welcomed. No mushrooms springing from the forest soil in splendid autarchy; it is a different anthropology to that of Hobbes, one of the archpatriarchs most frequently quoted by feminists to denounce modern atomism and radical suspicion of the other.[10]

Reason is not conceived as a distanced elevation from which an accomplished self surveys and chooses the most attractive options. The second distinguishing factor for humans after their cultural tradition is the combination of 'reason and experience'. In Pannenberg's review of Herder, this feature contains the equally significant need for 'self-formation', of taking a stance towards the educational influences one has been exposed to and forging a critical appropriation. One can see the *Bildungsromane* of this period as exemplifying the critical and creative encounter of a person's unchosen conditions with the active shaping of his selfhood. The third factor Herder considers necessary in the process of becoming human is 'divine providence' in educating humanity. For him, it is the image of God in which each one is made that directs us on the path of attaining our specific version of humanity.[11] The most decisive difference from animals, the gift of language, only exists in multiple systems.

What the Romantic period took from the fact of manifold languages against the early Enlightenment's bent towards universal reason is that diversity is meant to be. The plural ways of encoding and deciphering, of suggesting and guessing meaning, call for translation skills, reinventing another tongue in one's own, and allowing for difference to be left unharmonized. The *Origin of Language* makes it clear that perfection can only exist through diversity. The resources which the 'non-colonialist' attitude resulting from this evaluation offers to current discourses of multiculturalism have been noted.[12]

Do these idyllic descriptions have anything to say to today's planners of human development? First, Herder's critique offers similar insights to Siep's. Being learners in communities and participants of tradition, we have to take the conditions of our particularity seriously. Perfection as such is an abstraction from the all-decisive specificity of one's situation. With this insight, bored by a superficial concept of a generalized rationality, the Romantic authors went beyond the withering critique of the optimism of progress in this 'best of all possible worlds' that Voltaire expressed in *Candide* (1759)[13] more than a decade before Herder's *Origin of Language* (German original 1772).

10   S. Benhabib, *Situating the Self: Gender, Community and Postmodernism in Contemporary Ethics* (Cambridge: Polity Press, 1992), p. 156.

11   Pannenberg stresses how Herder regained a view of human life that comprised but also went beyond morality. The active participation of humans in God's gracious action takes place on more than the moral level. 'Herder's recourse to the idea of the divine image seems thus to be an expression of his opposition to the idea of a human self-fulfillment through active self-enhancement. In order to realize their human destiny, their humanity, human beings remain dependent on the most varied influences from outside and on the harmonious contribution of these to the advancement of their humanity. Their disposition to be like God is therefore fulfilled only by God himself, through the operation of his providence... The concept of the image of God is here not replaceable by anything else.' Pannenberg, *Anthropology*, p. 53.

12   N. Smyth, *Charles Taylor: Meaning, Morals, and Modernity* (Cambridge: Polity Press, 2002), p. 154.

13   Voltaire, *Candide, or, Optimism*, trans. and ed. Theo Cuffe (London: Penguin, 2005).

Secondly, Siep's appreciation that obstacles are part of a person's learning process and sometimes windows for new discoveries, not something necessarily at odds with one's fulfilment, finds its precedent in Herder's view that the task of becoming human involves working with what is given, and of being receptive and open to possibilities of meaning existing prior to one's agency. However, while admitting the advantages of an attitude open to being enriched, compensated and surprised by learning and tradition, the counter-question could be raised: What is wrong with perfecting the genetic bases for future skills? Is it not possible to enjoy particularity and diversity on the one hand *and* improve individual possibilities on the other? Why should it be impossible to be in favour of both?

The answer to this 'both-and' position lies in what Charles Taylor considers to be Herder's lasting, if not uncontroversial, legacy: the integrity of the 'voice of one's self'. The idea behind this appreciation is the Enlightenment-critical view that each one is his own measure; the rationalist universality of one measure for all is mistaken.[14] Thus, the appreciation of diversity in the uniquely human gift of language leads to a normative emphasis on singularity which is inimical to comparative grading and thus to the mindset behind the goal of enhancing. Schleiermacher's distinction between individual and exemplar reflects this insight. Yet before turning to him, two more points need to be made. They put into question the perception that Romanticism is about the celebration of difference in harmonic ease. The first concerns the antithesis present in the self, the second the problem contained in the label 'expressivism', which leads back to a Spinozistic view of totality and its parts.

The following brief sketch of Herder's developing views on selfhood is to show how difficult it is to label a complex period of rapidly changing self-interpretations; it can also put into perspective the typical fears of modern selves at the beginning of the 21st century by comparing them with those analysed at the beginning of modernity, and with the solutions suggested then.

ii. The tension between the finite and the infinite

The recognition of diversity and of the incommensurability of each individual's measure may be a spirited response to Rationalism's insistence on the tradition-sweeping authority of general reason, but it offers as much of a problem as of a solution. What exercised Herder's and his contemporaries' imagination were the divisions within each original individual. The very sequence of proposals from the same authors testify to their own self-critique and torn state. If one thing becomes clear in comparing theirs to current analyses it is that enhanced memories or more able physiques will not solve the depth of conflict inherent in human subjectivity. There are no technical fixes available for them; while the variety of measures of fulfilment, the

---

14   Charles Taylor, *Sources of the Self: The Making of the Modern Identity* (Cambridge: Cambridge University Press, 1989), p. 369. On p. 375, Charles Taylor summarizes and quotes Herder: 'The differences are not just unimportant variations within the same basic human nature... Rather they entail that each one of us has an original path which we ought to tread; they lay the obligation on each of us to live up to our originality. Herder formulated this idea in a telling image: ... "Each human being has his own measure, as it were an accord peculiar to him of all his feelings to each other"... Each person is to be measured by a different yardstick, one which is properly his or her own.'

playfulness and irony may appeal to us, it was the Romantics themselves who realized that no shortcuts were possible in trying to reconcile the human consciousness of their finitude with their striving for the infinite.

The stage at which Charles Taylor leaves the portrayal of Herder with whom he sees the 'expressivist turn' find 'its first important articulation'[15] is exactly the point where he radically reverses his earlier diagnosis and therapy. The immediacy of feeling all at one with the totality of nature is left behind as an immature and morally suspect intuition. The earlier stage of 'giving a central and positive place to sentiment in the moral life' is now superseded by an appeal to reason to counterbalance its overbearing enthusiasm. That for 'Herder, all passions and sensations "can and must be operative, precisely in the highest knowledge"' (371–72) is now passé, as is the 'perfect fusion of the sensual and the spiritual' which 'tends to dissolve the distinction between the ethical and the aesthetic', and for Taylor foreshadows 'the "flower generation" of the 1960s' (373). By emphasizing the tension in selfhood given in the ongoing struggle between objectifying reason (*Verstand*) and heart, Herder dismisses his earlier departure into the mysticism of the 'all-one'. It is now the idea of difference that is accepted as necessary to gain selfhood. His earlier one-sided embrace of relationality in being part of the whole is now suspected of being self-aggrandisement in the form of an indeterminate reverence for totality which cannot be distinguished from worshipping the ego. This is why parents' love of their children, *Elternliebe*, is now seen as the highest form of love. Not erotic love, nor friendship qualify for this rank, because they are too fusion- and exchange-centred as opposed to being able to give birth to and sustain lasting otherness.

> The love of parents is *divine*, because is it not self-centered and very often thankless. It is *heavenly* because it is able to divide itself into many and still remain whole, always undivided and without jealousy. Finally, it is also *eternal* and *infinite*, because it overcomes love and death…[16]

Herder seeks the traces of being created in the image of God in essential traits of human relations and sees human perfection as a self-formation towards this image that continues to depend on God's providence. At the same time, the sense of deep tension within the self shows his proposal in much greater proximity to the Protestant self-experience of the *simul iustus ac peccator*[17] than to the flower generation's rehabilitation of the senses.

The source of 'expressivism'
Yet, the lasting influence of Spinoza also in his second reception after Herder had distanced himself from the idea of a dissolution of the individual in the totality of the whole remains unaccounted for in both Pannenberg's and Taylor's portraits. Romantic 'expressivism', however, Taylor's key phrase for the period and for its

---

15   Taylor, *Sources*, p. 368. Further page numbers in the text.

16   Cf. H. Timm, *Gott und die Freiheit. Studien zur Religionsphilosophie der Goethezeit. Bd. I. Die Spinozarenaissance* (Frankfurt: Klostermann, 1974), pp. 282–99, original italics. The Herder quote on parents' love of their children is from p. 296.

17   Cf. Timm, *Gott und die Freiheit*, pp. 285–86, with reference to the influence of Hamann's Lutheranism.

lasting influence on current self-understandings, and the corresponding concept of 'nature as source'[18] cannot be understood without this background. Even the description of the 'heavenly' ability of parental love 'to divide itself into many and still remain whole' carries Pantheistic overtones. Whether this points to Herder identifying 'God' with Spinoza's *Deus sive natura*, or whether his reconstruction of Spinoza's theory of divine attributes as 'substantial forces' remains theistic, is controversial. His insistence in his debate with Hemsterhuis that the individual maintains his or her distance even as the appearance of the self-limiting 'contraction of God'[19] is shared by Schleiermacher. Each person has her own intuition of the universe of which she is 'one irreplaceable presentation'.[20] Already in the 'Speeches' (1799) a double structure of human striving is proposed, one towards attracting and incorporating what is other into one's being, the other towards expanding and communicatively expressing one's interior self.

Just how much the tension between the finite and the infinite has been radicalized is manifest in the Christology of the Speeches where Christ's death on the cross is reinterpreted speculatively as the breaking point to which the appearance of the absolute in the finite is doomed. Some of the author's own revisions in the later editions and his new departure in the two versions of the *Glaubenslehre* admit to the questionability of this theological reconstruction of Christianity where 'corruption and redemption, enmity and mediation are two sides of this intuition that are inseparably bound to each other'.[21] Yet what its extreme form does convey is the deep longing that the religious experience of God's withdrawal from the finite evokes. The dismay at the opposition between the finite and the infinite leads to defining Christianity in the character of its feelings as an 'unsatisfied longing' and 'holy sadness' (*heilige Wehmut*).

The conclusions to be drawn from Herder's and Schleiermacher's Romantic thinking for the project of genetic perfection can already be stated:

- Whatever perfection is to be achieved, it cannot be at the price of diversity.
- The roots of human imperfection lie deeper than the level of biology.[22]

---

18 Cf. Taylor, *Sources*, p. 374: 'My claim is that the idea of nature as an intrinsic source goes along with an expressive view of human life. Fulfilling my nature means espousing the inner *élan*, the voice or impulse.'

19 Timm, *Gott und die Freiheit*, p. 302: The relationship of the finite modes of God after God's individualization is described as '*inesse* and *exprimere*, inwardness and expression that became basic concepts of Romanticism, hermeneutics and the modern philosophy of life. Initially they were Spinozisms.' P. Grove questions Timm's interpretation of Herder as Spinozist in '"Vereinigungsphilosophie" beim frühen Schleiermacher und bei Herder', in U. Barth and K.-D. Osthövener (eds.), *200 Jahre, Reden über die Religion* (Berlin and New York: W. de Gruyter, 2000), pp. 328–43 (340–41).

20 Grove, '"Vereinigungsphilosophie"', p. 342.

21 I have discussed the changes in 'Die "Anschauung des Universums...zur Vollkommenheit ausgebildet". Zur Christologie der "Reden"', in Barth and Osthövener (eds.), *200 Jahre, Reden*, pp. 714–28. This and the following quote are from Schleiermacher, *On Religion: Speeches to its Cultured Despisers*, intro., trans. and notes R. Crouter (Cambridge: Cambridge University Press, 1988), pp. 213–17.

22 Taylor comments on such naturalism in *Sources*, p. 106: Those 'contemporary modes of

- That every person as a finite representation of the infinite is his own measure does not mean incommunicability but that individuals are there to enrich, not to compete with each other.

The significance of two further thoughts the theologian Schleiermacher shares with the philosopher Herder, those of an underlying source of expressions, and of each individual's incontrovertible finiteness, have to be developed from the mature, post-Romantic Christology of the *Glaubenslehre*.

b. *Potency versus Finite Expression, Fulfilment versus Bad Infinity (F.D.E. Schleiermacher)*

The first of the two principal insights that continue to be incisive in an age aspiring to genetic perfection is stated in paragraph 93.2 of the *Glaubenslehre*, where empirical perfection in all elements of Christ's self-expression is declared as Christologically beside the point. What matters are not finite, contextual expressions but the power, 'potency' or 'capacity' underlying them.

i. *Perfection in the capacity, not in the finite manifestations, of Christ's God-consciousness*

The point of Schleiermacher's argumentation is an anti-docetic and anti-supranaturalist understanding of Christ's unsurpassable position; yet what he explains with regard to this unique mediator uniting divinity and humanity in his person is valid as well at the level of mere human qualities.

> [We are] not concerned with the multifarious relationships of human life – as though Christ must have been ideal for all knowledge or all art and skill which have been developed in human society – but only with the capacity (*Kräftigkeit*) of the God-consciousness to give the impulse to all life's experiences and to determine them... There are those who say it is not only possible but our duty to go beyond much of what Christ taught to his disciples, because He Himself (since human thought is impossible without words) was seriously hindered by the imperfection of language from giving real expression to the innermost content of His spiritual being in clearly defined thoughts; and the same, it may be held, is true in another sense of His actions also, in which the relations by which they were determined, and therefore imperfection, are always reflected. This, however, does not prevent us from attributing to Him absolute ideality [*Urbildlichkeit*, being the archetype of God-consciousness] in His inner being, in the sense that that inner being may always transcend its manifestation, and what is manifested be only an ever more perfect presentation of it.[23]

What Schleiermacher discovers with regard to Jesus Christ's distinctive God-consciousness, the difference between expression and source, offers a critical insight regarding the goal of genetic perfection. There is no point in trying to enhance individual empirical features; if anything, one would have to enhance the source, the

---

thought...are quite unaware of the way in which our modern sense of the self is bound up with and depends on what one can call a "moral topography". They tend to think that we have selves the way we have hearts and livers, as an interpretation-free given.'

23   Schleiermacher, *The Christian Faith*, ET of 2nd German edn 1830–31, ed. H.R. Mackintosh and J.S. Stewart (Edinburgh: T & T Clark, 1986), pp. 378–79.

powerhouse that fuels and integrates these abilities. For Schleiermacher, the source in the case of the singular person of Jesus Christ is the utter potency of his God-consciousness over his sensible self-consciousness.[24] This insistence implies two conclusions:

1   It challenges secular orders of goods by establishing piety as the goal of every possible perfection because the tension that marks human subjectivity can only be resolved, if at all, at the level of religion, not through knowledge and not even through morality. In this sense, Christianity has raised the stakes in the perfection debate.

2   Even if one disregards the Christological context and focuses only on the individuality at the source of empirical traits, the insight for genetic projects of enhancement is a cause for concern: by enhancing performance, one may destroy the source of competence, a person's irreplaceable individuality at its very centre which Schleiermacher calls 'immediate self-consciousness'. Then genetic measures of perfection turn out to be measures of genetic streamlining, of making something unique into something typical. Based on a structural contempt for the selfhood that is the source of all finite manifestations, intervention would optimize features but destroy originality and diversity. Here, Schleiermacher's insight joins Kierkegaard's concept of being able to be oneself that Habermas invoked as a limit to anyone's scope of choice for others.

The difficulty of being a self – marked inescapably by division, internal struggle and dynamics of conflict that create conditions of sin – which cannot be undercut by genetic reprogramming would be complicated even more by the intervention of others. Both the personal sense of responsibility – Habermas's concern – and of achievement would be undermined. Whose excellence is it, the laboratory's precision, the parents' clever choice ('it's you, but only the best of you', as the scientist in the film *Gattica* suggests), or that of the resulting optimized individual who decides to act true to his implanted genes? From Schleiermacher's concept of the irreplaceable self it becomes clear that the action of enhancing is wrong, while of course the result, the enhanced and even the cloned person, would be respected as a subject of human dignity and as a bearer of God-consciousness.

ii. Perfection as unachievable in the species
The second conclusion to be set against the evolutionary hopes of some proponents

24   Here, the theological discussion of the *Glaubenslehre* is ongoing: on the definition of piety in paragraphs 3 and 4, on the need for its theological insistence on the unity of God's decrees of creation and redemption in paragraph 48, on the relationship between the human person's creation in the image of God, the role of the Fall, and the understanding of redemption and of fulfilment, and whether Schleiermacher's definition of the unsurpassable dignity of the person of Jesus delivers as much as he thinks it does. See, e.g., my *Das Urbild des Gottesbewusstseins* (Berlin and New York: W. de Gruyter, 1990), pp. 121–31, 173–215. See also S. Schaefer, *Gottes Sein zur Welt* (Regensburg: Pustet, 2002). Despite these specific points of critique the enterprise of relating the truth of the Christian faith to the general consciousness of truth that philosophy investigates is to be defended as an ongoing task posed by the Christian heritage since the early Apologists in response to 1 Peter 3.15, to give account of the hope within us.

of enhancement for the human race is Schleiermacher's sobering analysis that *if* such a goal is to be reached, it has to happen in one individual. If it does not, the race cannot make up for this failure. Thus, if creation is to be fulfilled, it can only be achieved in one individual, Christ.

> [The hope] that some day the human race, if only in its noblest and best, will pass beyond Christ and leave Him behind...clearly marks the end of the Christian faith, which on the contrary knows no other way to a pure conception of the ideal than an ever-deepening understanding of Christ...as soon as we grant the possibility of a continued progress in the potency of the God-consciousness, while denying that its perfection exists anywhere, we can also no longer maintain that the creation of man has been or will be completed, since undoubtedly in progress thus continual perfection remains always only a bare possibility. And this would be to assert less of man than of other creatures – for it may be said of all more limited kinds of being that their concept is perfectly realised in the totality of individuals, which complete each other. But this cannot hold of a species which develops itself freely, if the perfection of an essential vital function be posited in the concept but actually found in no individual; for perfection cannot be obtained by adding together things that are imperfect.[25]

Does this argumentation amount to 'bottomless speculation' (W. Dilthey), or is it the one decisive argument first against Feuerbach and Marx, and now against Nietzsche, Sloterdijk and contemporary hopes for a new evolutionary breakthrough? As to the possibility of the species achieving perfection, Schleiermacher's reflection clarifies that it either has to be there in one person or it never will be. Adding finite entities to one another will only ever result in bad infinity but not in a qualitative change-over to perfection.[26]

Conclusion

The call to achieve genetic perfection by enhancing one's children's genes may seem plausible in a relentlessly antagonistic, begrudging and power-accumulating culture where physical, ecological and personal resources are getting scarce. Driven by the fear of losing out, our age can benefit from testing earlier responses to the break-through mentality of some proponents of the Enlightenment. Well before spelling out his view of the practice of education within asymmetric relationships in his pedagogical writings, the early Romantic Schleiermacher urged in his 'Outline for a Reasonable Catechism for Noble Women' (1798): '5. Honour the individuality and free will of your children so they may prosper and live heartily on the earth.'[27] (*Ehre die Eigenthümlichkeit und Willkür Deiner Kinder auf dass sie kräftig leben auf Erden.*) As a rewriting of Moses' rendering of the divine

25   Schleiermacher, *The Christian Faith*, pp. 378–79.

26   I have developed this point in greater detail in *Urbild*, pp. 178–80. A short English version of the book's argument is contained in my 'The Transcendental Turn: Shifts in Argumentation between the First and the Second Editions of the Glaubenslehre', in *New Athenaeum/Neues Athenaeum* (Lewiston, NY: Edwin Mellen Press, 1992), pp. 21–41.

27   Besides the translation (p. 60), Ruth Richardson provides a detailed comment and historical evaluation in R.D. Richardson, *The Role of Women in the Life and Thought of the Early Schleiermacher (1768–1806)* (Lewiston: Edwin Mellen Press, 1991).

command to 'Honour your parents', it denotes a shift in emphasis from what is due to one's parents, to the fulfilment they can find if they delight in the 'wilfulness' of their children.

Ideals of perfection have come a long way. For Aristotle, children could not be happy because they lacked wisdom, *phronesis*, intellectual insight.[28] In Jesus' empowering re-evaluation, they symbolized not just their parents' and their people's future. Jesus valued children in themselves and saw their own existing, not just nascent or developing, perfection in their utter trustfulness. For Jesus, the reign of God had already begun, and the voices of children were a sign of it.

Unlike John the Baptist, Jesus's creation spirituality did not need the overturning of the present state of inherent imperfection and deep flaw, in relation to the exigencies of society. The 'new person' he asks for is modelled on a different measure of perfection. This is why Luke renders the idea that Matthew expressed as 'perfect' (*teleios*) at the end of the Sermon on the Mount with a different emphasis: 'Be merciful as your father in heaven is merciful' (Lk. 6.36).

Re-imagining perfection as striving to be merciful would give rise to quite different centres of excellence than the ones envisioned in the current demands of economic globalization. Their benchmark would be inclusivity. It would turn back the gaze from manifestly achievable goals to the inner force of striving. The agent and the content of perfection then is a person's God-consciousness. If there had been any doubt, it now becomes clear that a New Testament quote such as the following is not a manifesto, but a critique of perfectibility: 'This is the Christ we are proclaiming, admonishing and instructing everyone in all wisdom, to make everyone perfect in Christ' (Col. 1.28). And we are lucky to have one version of Matthew's call to perfection which opts for this focus, the Jerusalem Bible taking the liberty of translating *teleios* as follows: 'You must therefore set no bounds to your love, just as your heavenly father sets no bounds to his' (Mt. 5.48).

---

28 Cf. J. Rohls, *Geschichte der Ethik* (Tübingen: Mohr-Siebeck, 1991), p. 66.

Chapter 12

FUTURE PERFECT? GOD, THE TRANSHUMAN FUTURE AND THE QUEST FOR IMMORTALITY

Celia Deane-Drummond

1. *Introduction*

Nanotechnology, microchip technology and genetic engineering all serve to conjure up images of the cyborg, a mixing of human and machine leading to a human-like creature that has elevated functions over and above that normally associated with material, finite bodies. Transhuman[1] philosophers have urged a greater acceptance of such technologies for the sake of what they perceive is both the perfectability of human attributes, but also a reaching beyond this towards breaking down the limitations of finitude, a quest for a secularized immortality.

How far are such technological developments realistic from a medical point of view? If ageing is seen in medical terms as a 'disease' that has to be 'cured', has the quest for immortality also crept into medical practice, albeit in a seemingly more innocuous form? What are the ethical resources within medical bioethics that might serve to foster or resist such developments? What are the particular virtues that growth in the Christian life would encourage, for example, growth in wisdom, prudence, fortitude and temperance? More broadly, how far are such developments ethically acceptable in terms of justice as virtue and the common good, in the light of human responsibility to future generations?

This chapter will ask what might be the underlying motives and content of such immortality and offer a critical analysis in the light of Christian theological views about human perfection, finitude, eschatology and immortality. I will argue that theological reflection allows a deeper perception of, first, the place of perfection in the human life and, secondly, an acknowledgement of mortality as a good, one that accepts death, while being compassionate towards those who suffer. I will also suggest that where individual choice and consent are elevated as ethical

---

1 Transhumanity is more commonly understood as an alliance of humanity with technology for the specific intent of allowing human capacities to reach far beyond any 'natural' limitations, set only by the technology as such. Such enhancement technologies might include the use of drugs, nanotechnology, genetic engineering, artificial implants and so on. A case for transhumanity in relation to genetic engineering, for example, is found in N. Bostrom, 'Human Genetic Enhancement: A Transhuman Perspective', *The Journal of Value Inquiry* 37.4 (2003), pp. 493–506. For a more detailed discussion of the similarities and differences between posthumanity and transhumanity, see Ted Peters, this volume.

norms, the quest for immortality becomes privatized according to the norms set by liberal democratic views on justice.

2. *Technology and the Holy Grail*

The drive for perfection can hardly be thought of as a new phenomenon. Alchemy, the precursor of modern chemistry, adhered to a vision of perfection when sophisticated medical technologies had not yet emerged. Alchemists believed that by close imitation of nature eventually that elusive 'Fountain of Youth' would come their way, conjuring their followers to consume gold in the mistaken belief that it held anti-ageing properties.[2] The thirteenth-century writer Roger Bacon, often thought to be the father of the pro-longevity movement, believed that a human lifespan could be restored to 900 years.[3] Francis Bacon, writing in the seventeenth century, argued for a more experimental approach to science; technology, he believed, should be directed towards the service of humanity, as the application of science was its greatest achievement.[4] The models that he used for technology came from within the natural world, so that technology could be inspired by nature, but also reach beyond it in order for humanity to control natural processes. Aristotle believed that the regularities in nature imitated art, but Bacon was more interested in those occasions where nature produced wonders, so that 'Whereas for Aristotle monsters and other errors of nature destroyed order, for Bacon they created and inspired new orderings of things.'[5] Descartes believed that automata made by human hands were similar to natural kinds made by God; the difference was simply a matter of scale and complexity. But this close alliance between nature and art was soon to fade with the invention of the microscope. Robert Hooke found that the point of the finest needle was as ragged as a mountain range.[6] The separation of the natural art of God and that created artificially by humanity seemed irreconcilable.

Yet in today's culture we are once more becoming reawakened to that dream of imitating nature. Indeed, the desire is not simply to imitate her, but to improve on her 'natural' capabilities. No longer is God considered the author of that nature; it simply confronts us through the meanderings of evolutionary history. Of course, knowing humanity has taken so long to arrive on the planet after so many millennia of cosmic and evolutionary experiment gives some biologists reason to doubt the wisdom of too much tinkering. But transhuman philosophers are more cavalier. They are aware, for example, of the possibility of genetic enhancement through the addition of genes that give extra capabilities. They are also aware that

---

2   J. Hedley Brooke and Geoffrey Cantor, *Reconstructing Nature* (Edinburgh: T & T Clark, 1998), p. 316.

3   S. Jay Olshansky and Bruce A. Carnes, *The Quest for Immortality* (New York: W.W. Norton, 2001), p. 40.

4   Lorraine Daston and Katherine Park, *Wonders and the Order of Nature* (New York: Zone Books, 1998), p. 291.

5   Daston and Park, *Wonders and the Order of Nature*, p. 291.

6   Daston and Park, *Wonders and the Order of Nature*, p. 300.

nanotechnologies, by imitating nature on a molecular scale, come much closer to the possibility of the fusion between nature and art.

In medicine nanotechnology is already being used to deliver drugs in trials with an accuracy that boggles the imagination.[7] Cosmetic surgery is also becoming much more widespread in the Western world, available on demand to those who have the financial resources. All these technologies add fuel to the dream of human perfectability. Anti-ageing technologies, which have always been an aspect of cultural and scientific interest, take on a new impetus in the light of the possibility of irreversible genetic enhancements. Sheila Rothman and David Rothman, for example, claim that 'once the genetics of ageing are understood, longevity might increase not merely by ten or twenty years, which would be the result of curing today's leading causes of death, but by seventy to ninety years, with an average life span of 140 to 160 years'.[8] They reject the ideal of the natural, citing examples to show how much medical practice, such as organ transplantation, is also unnatural, but not subject to the same objections. The reordering of nature also has its own historical precedents and parallels in manipulation of non-human nature.[9] They also suggest that the role of doctors has shifted; they are no longer simply called upon to cure disease, but rather respond to patients' psychological feelings of unhappiness related to fantasies of perfection. In such scenarios the risks of particular treatments are often not sufficiently aired. What is particularly interesting about their analysis is how they have shown that what first started as a particular treatment for a given disease, later became the means for securing particular enhancement therapies.

A good historical example of the shift towards enhancement therapy from original intentions to cure disease is the use of oestrogen therapies.[10] This also connects with the cultural pressures for women to be concerned about their bodies in stereotypical images of beauty. Oestrogen was originally used in the 1920s and 1930s, and given to women suffering from loss of monthly periods due to hormone imbalance, that is endocrinopathic amenorrhea. The treatment caused pseudo-menstruation that was unrelated to ovarian function, but doctors continued to prescribe on the basis that it had beneficial psychological impact. It was also used for treating symptoms of the menopause. Later, oestrogen was sold on the basis of its re-invigorative properties, in spite of known carcinogenic properties from experiments on rodents in the late 1930s.

In the late 1950s gynaecologists, notably William Masters, started recommending lifelong use of oestrogen in spite of its risks, using emotionally charged language that demeaned those women passing through the menopause, such as

7   David A. LaVan, Terry McGuire and Robert Langer, 'Small Scale Systems for In Vivo Drug Delivery', *Nature Biotechnology* 21 (2003), pp. 1184–91.

8   S.M. Rothman and D.J. Rothman, *The Pursuit of Perfection: The Problems and Perils of Medical Enhancement* (New York: Pantheon Books, 2004), p. ix.

9   For further discussion of the historical, social, cultural, theological and ethical aspects of manipulating the natural world, see C. Deane-Drummond and B. Szerszynski, *Re-Ordering Nature: Theology, Society and the New Genetics* (London: T & T Clark, 2003).

10   Full discussion can be found in Rothman and Rothman, *The Pursuit of Perfection*, pp. 43–95.

that they were victims of 'steroid starvation' that rendered them 'neutral' in terms of gender. The ageing process for women was couched in terms of sub-optimal function, including those diseases related to shifts in endocrine levels, such as osteoporosis. At this stage cardiac disease and dementia were also connected with low oestrogen. Masters also dismissed the possibility of cancer risk by suggesting that only reckless use would cause a problem. The growth in the use of oestrogen in the United States illustrates the influence of this kind of rhetoric. In 1958 there were one and a half million prescriptions given; by 1975 this had risen to 27 million. Once the contraceptive pill became more established, women were encouraged to stay on oestrogen 'from adolescence through to senescence'.[11]

In the 1960s Robert A. Wilson, a Brooklyn gynaecologist, received a large grant from three pharmaceutical companies in order to set up a foundation to promote the use of oestrogen for anti-ageing. Alongside its medical benefits, women were portrayed as more attractive, vibrant and desirable when on such treatments. His book *Feminine Forever* carried an extensive bibliography, but the research was weak, 'in fact it was a collection of personal observations and anecdotes'.[12] According to him, menopause was nature's way of 'robbing women of their womanhood', implying that women also had a 'patriotic duty' to avoid such effects as they would not be able to serve their community when burdened by such effects of the menopause. There was little scientific research on appropriate dosage; his recommendations for dosage were based on guesswork unrelated to research.

In the 1970s other medical practitioners began to offer a critique of this approach, suggesting that menopause was not pathological; rather, after a period of imbalance, adjustments were possible. Slowly more evidence accumulated for the link between the risk of cancers and use of oestrogen in women. Eventually the Federal Drug Authority advised against its use, but this was met initially by strong objections from the American College of Obstetrics and Gynaecology. Many had a stake in promoting the benefits of hormone replacement therapy (HRT) as grants were given to support clinical research in the field. In 1993 a large-scale trial on use of HRT was conducted, where, in spite of slowly accumulating evidence to the contrary, sponsors anticipated that the benefits of HRT would be forthcoming. Over a year 10,000 women taking HRT when compared with those women who did not have treatment had seven more coronary heart diseases, eight more strokes, eight more blood clots, eight more invasive breast cancers, six fewer colorectal cancers and five fewer hip fractures. The trial was stopped immediately as a result, but some still persisted in arguing for the benefits of HRT, suggesting that it might be used for particular groups of women.

The above case study is instructive as it shows the interplay between cultural expectations of women, medical therapies, and the slip towards enhancement technologies under the guise of medical practice. It also shows the extent to which medical practitioners are vulnerable to the temptations of research grants from companies that have particular vested interests in promoting particular drugs or

11   Rothman and Rothman, *The Pursuit of Perfection*, p. 75.
12   Rothman and Rothman, *The Pursuit of Perfection*, p. 76.

treatments. This led to clouded judgements about the benefits of the technology and also a shift in the role of medical practitioners, as well as the expectations placed on them by those caught up in the cultural drive for perfectibility. It also shows the cultural history of anti-ageing technology in particular is highly pervasive, being expressed in the last century in terms of hormone replacements, whereas the shift in the twenty-first century is more likely to be in terms of the use of genetic enhancements to achieve anti-ageing effects. Such a shift opens up a new dimension to anti-ageing, for it becomes possible to envisage a significant extension of life, not merely an enhancement in perceived quality of life.

Where does all this fit in with what the sociologist Zygmunt Bauman has called life strategies? He believes that our culture has two facets. The first activity of culture relates to survival in the face of awareness of mortality, 'pushing back the moment of death, extending the life span, increasing life expectation'.[13] The second activity of culture 'relates to immortality – surviving, so to speak, beyond death, denying the moment of death its final say, and thus taking off some of its sinister and horrifying significance'.[14] But immortality is, for him, not simply the absence of death; rather, it is 'a defiance and denial of death', so that 'There would be no immortality without mortality. Without mortality no history, no culture, no humanity.'[15]

The implications of this view for the present context are striking. The first concern of human culture identified by Bauman, that is, human concern for *survival*, is related to the quest to deconstruct mortality that is inherent in modernity. It could, perhaps, be thought of as an extension of the never-ending battle against particular diseases and threats to life; only in this case the battle against particular diseases associated with ageing is included in this category. When medical practitioners place ageing-related diseases within the list of those problems that are potentially soluble, they become manageable, a project that seems acceptable within current medical practice. There is, of course, a more sinister aspect of this dream, for 'All too often, and certainly much more often for moral comfort and political placidity, the audacious dream of killing death turns into the practice of killing people.'[16] In other words, a mortality-denying, imperfection-denying culture that is implicit in much of the modern project will not be able to accept mortality and imperfection as such, expressed paradoxically through insidious policies of eugenics and euthanasia. Bauman believes that a lack of sense of mortality is culturally charged, for its expression in such policies undermines the very fabric of history and culture.

The second facet of Bauman's analysis relevant in this context is related to a *denial* of death and the drive for immortality. But this is paradoxically set in a context of modernity where mortality is also being denied. I am implying, in other words, that enhancement technologies are more than they claim to be, simple

---

13 Zygmunt Bauman, *Mortality, Immortality and Other Life Strategies* (Cambridge: Polity Press, 1992), p. 5.
14 Bauman, *Mortality, Immortality and Other Life Strategies*, p. 6.
15 Bauman, *Mortality, Immortality and Other Life Strategies*, p. 7.
16 Bauman, *Mortality, Immortality and Other Life Strategies*, p. 160.

improvements in the quality of life and human existence. What, might one ask, is improper about such desires? Historically speaking, one could argue that in the nineteenth century the thought of living to a hundred years might have seemed perverse, where life expectancy was much lower than it is today. Yet in one sense enhancements could be thought of as outside the usual drive for medical practice that has sought to address issues of mortality, that is human *survival* and therapies related to this. In the nineteenth century the survival concerns would have been uppermost, though arguments for increasing medical interest in the quality of life are also becoming more vocal in contemporary society, not just in consideration of patients who are terminally ill, but also throughout life. The growth of so-called alternative medicine is a testimony to this shift in Western cultures. But I suggest that in addition to these motivations there is interlaced within enhancement technologies a desire for perfection that is more properly conceived in the context of cultural drives for immortality. Such drives eventually find expression in the quest to counter even the moment of death, linked in a more obvious way to immortal goals.

Has transhumanity also been influenced by postmodernity? For postmoderns there is no 'life plan', and in this, the quest for immortality is undermined. Human identity is only momentary, linked with the present rather than the past or the future. Here 'the distinction between the mundane and the eternal, transient and durable, mortal and immortal is all but effaced'.[17] This leads to maximum scepticism about any long-term life plan, acknowledging that the skills learnt today will no longer be of any use in tomorrow's technology. In this way Bauman suggests that 'now in the wake of the distinctive job performed by modernity it is immortality that has been "tamed" – no more an object of desire, distant and alluring; no more the remote and high-handed God, commanding ascesis, self-immolation and self-sacrifice'.[18] In such a context he believes that the last shelter of reassurance is community. But the paradoxical outcome is that the work of modernity in removing mortality is now being undone, 'Death is back', so that even immortality comes under its spell.

My suggestion is that while many medical scientists are still working under a focused concern for survival needs, that is a life strategy of the conquest of mortality, subliminal postmodern cultural drives that undermine the distinctions between mortality and immortality also come to the surface. In other words perfection is still sought, but it is sought haphazardly in a way that is detached from specific goals. Now the Holy Grail takes on different forms depending on the moment; it is a nomadic search, sensitive to whatever is on the horizon at the time, but without the commitment of a lifelong task. Hence, the 'price of exorcising the spectre of mortality proved to be a collective incapacity to construct life as reality, to take life seriously'.[19]

17 Bauman, *Mortality, Immortality and Other Life Strategies*, p. 167.
18 Bauman, *Mortality, Immortality and Other Life Strategies*, p. 169.
19 Bauman, *Mortality, Immortality and Other Life Strategies*, p. 199.

3. *Theology, Eschatology and Perfection*

How might theology respond to these developments? Certainly, a secular eschatology that seeks to claim immortality for itself, while paradoxically undermining any basis for it through *eit_____ __ _____ denial of mortality in modernity *or* mortality's fusion with _____ inevitably fail to satisfy the human desire _____ correctly identified as being what cultur___ _____ in his assessment, claiming that 'medical an_ ___ _____ of human life has nothing whatever to do _ _____ il life" for eternity is not endlessness'.[20] The _____ ieological tradition human desire for perfec____ _____ human bodies and more on perfection in h____ _____ t has also admitted that perfection and imn_____ this world we have to be content with impe_____ th century there has been a revival of esch_____ some sense attuned to what will happen in the future, a____ _____ ie reign of God.[21] In particular, theologians have become critical of the future understood as simply an extension of the past that by nature is conservative; instead the new has its origin in the eternal God.[22]

[handwritten note:] 'Eternal life' has been misunderstood. ↳ Does not mean endless ↳ Not associated with body, but virtue).

How are we to describe that continuity and discontinuity that Christian faith affirms at the moment of death? A traditional belief in the soul that somehow flies away from the body at the moment of death no longer seems tenable. There are two other theological alternatives offered here; one is that God recreates the identity of the dead, the other is that the dead live on in the memory of God, a concept that 'appears to be able to find greater spontaneous acceptance in a computer age'.[23] Antje Jackelén argues, instead, for a third alternative, that the self is preserved in relation to God, not so much through self-conservation, as self-reception. Hence:

> only in this way is it completely possible to come to oneself and to find oneself; and indeed, eternal life then presumably has to do with finding more than oneself. Identity thus becomes a question of relation. If death and eternal life are described as that through which one's life or self finds itself eternally, then the basic assertion that God is love is violated, for love is inconceivable without alterity, without a dynamic giving and receiving. [24]

---

20   J. Moltmann, *In the End – the Beginning* (London: SCM Press, 2004), p. 152.

21   Jürgen Moltmann has pioneered the renewal of eschatology in Christian theology, as evidenced by a number of publications. See, in particular, his earlier *Theology of Hope* (London: SCM Press, 1967), *The Coming of God* (London: SCM Press, 1996), and also his more recent *Science and Wisdom* (London: SCM Press, 2003) and *In the End – the Beginning*.

22   Antje Jackelén, *Time and Eternity* (Philadelphia: Templeton Foundation Press, 2005), p. 213.

23   Jackelén, *Time and Eternity*, p. 217. This would also be consistent with transhuman dreams, though transhumanity might view survival of information as being sufficient to define immortality.

24   Jackelén, *Time and Eternity*, p. 218.

According to this view, the classical Platonic belief in the continuity of the soul that assumes that beyond death there is strict continuity with selfhood in this life is flawed, hence undermining another basis for those forms of secularist eschatology influenced by modernity that cling ever more limpet-like to present existence.

Nonetheless, I suggest that for Christian eschatology to be meaningful there needs to be *some* sense of continuity with this life in terms of experiences in the next. Jürgen Moltmann suggests that human biographical history is caught up at the moment of death with God, but death does not limit human relationship with God.[25] He argues that the contrast needs to be drawn between love/livingness and death, rather than an immortal/detached soul and a mortal body. In addition, continuity implies affirmation, for eschatology that is detached too far from material existence has negative implications for the way humanity perceives its relationship with the earth. Moltmann's starting point for eschatology is Christological, that is, rooted in the death and resurrection of Christ, and this has cosmic significance for the whole of creation.[26] Ernst Conradie argues, similarly, that the resurrection of Christ has significance not just for human existence, but the whole of material creation as well.[27] The resurrection of Christ cannot be regarded simply as a resuscitation of a corpse, nor a return to this form of life, but rather the first fruits of the transformation of the whole of creation. Christianity also admits to faith in some form of *material* resurrection, even if the bodily life is not identical with life on earth as currently experienced.

One of the arguments of this chapter is that there has been a distortion in our sense of mortality and immortality; while the former is denied through constant striving against death in the modern project, the latter is caught up with grandiose dreams about perfection as that relates to present existence. The future, in such a scenario, is simply in continuity with the present, rather than a more radical claim that the future meets us from ahead, which is embedded in Moltmann's understanding of eschatology. Recognition that all is not right with humanity might imply that the primary problem of human existence is not so much finitude but human sinfulness, not so much participation in the cycles of life, but attempts to escape that flux through our own human efforts.[28] Ernst Conradie points to the fact that millions of people do not have the luxury to worry about old age, as their lives are interrupted by violent and untimely deaths. He suggests that:

> millions of people die a premature and violent death through war, murder, rape, starvation, tragic accidents or deadly diseases (AIDS!). A purely 'natural death' is indeed rare, especially in Africa. Death, including our own death, as it actually meets us is inseparable from God's appropriate judgment on human sin. [29]

25 Moltmann, *In the End – the Beginning*, pp. 104–108.
26 See, for example, J. Moltmann, *The Way of Jesus Christ: Christology in Messianic Dimensions* (London: SCM Press, 1990); Moltmann, *The Coming of God.*
27 Ernst Conradie, *An Ecological Christian Anthropology: At Home on Earth?* (Basingstoke: Ashgate, 2005), pp. 68–70.
28 Conradie, *Ecological Christian Anthropology*, p. 157.
29 Conradie, *Ecological Christian Anthropology*, p. 157.

By this he seems to mean that the reality of death is bound up with shocking human sinfulness, evil and separation from God, so that in such circumstances restoration is impossible in this life. Although somewhat ambiguous from this citation, his view lacks any sense that such death is 'deserved' by those suffering under such sin. Such a realization of the tragedy of death in most contexts puts into perspective the drives for perfection and enhancement. Such drives avoid facing the tragic reality of a life cut off well before its prime, and the added injustices associated with uneven distribution of medical resources that make consideration of life extension and other enhancements the privilege of a relatively small minority, even if desired more widely.

Eschatology implies reconciliation and healing for human sin in a way that finds perfection in the next life, but can begin in this world. In other words, Christian hope includes the possibility that present negative attitudes and inordinate desires will be changed and challenged. Christian vocation looks not just to the future eschaton; but also to work for a better future here on earth. The question becomes, what is the appropriate focus in terms of alleviation of human suffering? Goals in medical science are commonly presented in terms of that alleviation, but it is worth noting what the nature of that suffering is, and how such a drive towards that alleviation impacts on others. In this I would disagree strongly with Rothman and Rothman's view that we do not need to consider fairness in the development of new technologies because, according to them, there are persistent inequalities in American health delivery that cannot be overcome.[30] Peter Vardy says something similar to Rothman and Rothman in suggesting that 'once the general principle of privilege is accepted, then there seems no good argument for preventing people buying genetic privilege for their children'.[31] This acceptance of a principle of privilege amounts to a passive acceptance of injustices in the face of inequalities of advantage and further denies the possibility of change, with the possibility of confronting both individual and structural sins.

Another theological stream that is worth mentioning in this context is the notion that eternity is in some sense woven into the mystery of all of life. This strand in theological thought puts more emphasis on the positive goal of living a resurrection life and serves to complement those strands more characteristic of Reformed Protestant theology that put more emphasis on human depravity, sin and corruption. God's presence is not simply that towards which humanity aspires in the future, but is also present with us now, participating in creaturely existence. Such experience brings hope that detracts from the horror of death and suffering, for God is present to both. This approach was common in the early history of the Church, reaching its climax in doctrines of *theosis*, the deification of humanity through participation in God.[32] Certainly, within the Eastern Patristic tradition,

---

30   Rothman and Rothman, *The Pursuit of Perfection*, p. xviii.

31   Peter Vardy, *Being Human: Fulfilling Genetic and Spiritual Potential* (London: Darton, Longman & Todd, 2003), p. 67.

32   One of the most prominent Eastern writers in this vein was the fourteenth-century theologian Gregory Palamas. See Gergios Mantzaridis, *The Deification of Man* (New York: St Vladimir's Seminary Press, 1984).

salvation was couched in terms of a 'pilgrimage into God' through theosis.[33] Such reflection stems from an interpretation of 2 Peter 1.4, where Christians are encouraged to be partakers in the divine nature. Theosis implies a union with God that is grounded in the dynamic indwelling of the Holy Spirit in communion with Christ. Moreover, according to Peter, a life lived in this way is expressed through the virtues. The bridge between an infinite God and creatures seeking God becomes expressed in, for example, Gregory of Nyssa's doctrine of *epektasis*, which suggests a constant reaching out after God.[34] It is also inappropriate to view such doctrines as contributing to a sense of elimination of individual identity through some sort of cosmic fusion with the Godhead, for participation in God implied a discovery of true human individual identity, rather than its obliteration. In addition, the implication is always that there is no possibility of union with God in this life, for the longing for God is never satisfied, 'there is simply a deeper and deeper penetration into darkness'.[35] Moreover, a mystical communication between Creator and creature – leading to an encounter that is properly personal – is assured rather than a transformation into divine substance (the latter is anyway a pantheist heresy).[36]

While more common in the tradition of the Eastern Orthodox Church, it is wrong to suppose that theosis is absent outside Byzantine theology, for it is also in the works of Aquinas.[37] The difference between Palamas, for example, and Aquinas is that the latter sees grace expressed through the theological virtues of faith, hope and charity. The ultimate happiness for Aquinas is participation in divine nature. Aquinas also believed that grace extended those possibilities that are inherent in humanity through its natural human capabilities. The extent of grace is in relation to the degree of charity that is given, the higher the charity the more the desire for God. It would be wrong to believe, therefore, that Aquinas considered perfection to be narrowly located in the intellectual capacities of the human person, given the prominence of charity in receiving grace.[38] Anna Williams summarizes the purpose of creation in Aquinas's thought as 'that all things become good and perfect and all things find their consummation in divine goodness, which is none other than the divine essence'.[39] The Eastern tradition, following Palamas, put more emphasis on perfection through divinization as attainable in this life. The Western tradition, on the other hand, was more

---

33   Rowan Williams, 'Macrina's Deathbed Revisited: Gregory of Nyssa on Mind and Passion', in Lionel Wickham and Caroline P. Bammel (eds.), *Christian Faith and Greek Philosophy in Late Antiquity* (Leiden: E.J. Brill, 1993), p. 244.

34   Andrew Louth, *The Origins of the Christian Mystical Tradition: From Plato to Denys* (Oxford: Oxford University Press, 1981), p. 89.

35   Louth, *Origins*, p. 89.

36   For a discussion of this point see Emil Bartos, *Deification in Eastern Orthodox Theology: An Evaluation and Critique of the Theology of Dumitru Stăniloae* (Carlisle: Paternoster, 1999), p. 46.

37   Anna N. Williams, *The Ground of Union: Deification in Aquinas and Palamas* (Oxford: Oxford University Press, 1999).

38   Williams, *The Ground of Union*, p. 39.

39   Williams, *The Ground of Union*, p. 66.

orientated towards the future, perfection viewed only as feasible finally in the next life. Both sought union with God as the ultimate goal. Both challenge the notion that human happiness or perfection can be achieved apart from God.

4. *Seeking Perfection through Attainment of Virtues*

I suggest that consideration of the virtues is particularly appropriate as bridges between more idealistic notions of theosis, as the ultimate goal of the Christian life, discussed above, and the practical day-to-day ethical decision-making that is necessary in order to inform practical policy-making and medical practice. In other words, the Christian vision for perfection needs to find its place in forms that are more concrete in expression, if it is to avoid the charge of gnosis, a denial of the world that detaches itself from material reality. Drawing on 2 Peter 1.4 and Philippians 3.13, 14, Gregory of Nyssa was explicit in making the link, claiming that 'whoever pursues true virtue participates in nothing other than God, because he is himself absolute virtue'.[40] Ethical perfection is, of course, impossible to attain, as there are no limits to this perpetual process of sanctification. It answers the charge: What difference might it make in practice to aim at perfection under-stood as participation in God and a radical discontinuity as well as continuity with the present expressed through theologically informed eschatology? In other words, while the secularist visions of eschatology can be critiqued through theo-logical analysis, if the outcome of such critique simply creates a vision that is in detachment from medical practice then it is all too easily safely ignored.

In the classic tradition, perfection was sought in this life through the develop-ment of particular habits or virtues. It is also worth noting that along with the specifically theological virtues of faith, hope and charity, Aquinas speaks of the virtue of wisdom, alongside the cardinal virtues of prudence, justice, fortitude and temperance. All of these virtues are capable of becoming grace-laden; that is, capable of being enhanced by the work of the Holy Spirit. Virtues are, in other words, ways of expressing in practical terms that aspect of Christian eschatology that is both now and not yet. It is now in the sense that such virtues are sought in this life and geared to make a difference to that life. It is not yet in the sense that such virtues are the beginnings of the breaking in of the future reign of God that expresses radical discontinuity with the present world. The ultimate goal and purpose of creation is, according to Moltmann, the glorification of God, so that 'Ethical existence is gathered up and perfected in the aesthetic existence of dox-ology.'[41] The virtues allow for an Aristotelian philosophy that recognizes the good possibilities in human nature, but for Aquinas it is still a nature that is transformed by the power of God's grace, which is closer to the Eastern tradition of participa-tion in God as the basis for attaining virtues. At the same time, the awareness of human sinfulness is present through a realistic acknowledgment of vices and the

---

40   Gregory of Nyssa, *The Life of Moses*, trans. A.J. Malherbe and E. Ferguson (New York: Paulist, 1978), 1.7, 31. I am grateful to Peter Heltzel for pointing to the explicit link between *theosis* and attainment of virtues in the work of Gregory of Nyssa.

41   Moltmann, *The Coming of God*, p. 324.

likely distortion in the virtues. Aquinas mediated successfully between the more Platonic emphases in the Augustinian tradition and a recovery of Aristotelian recognition of the importance of material, biological nature. Such mediation was achieved through the virtues that were both habits of mind that could be learnt, but also could be received as gifts in the context of the Christian community. In other words, Aquinas recognized elements of continuity with human nature, but also pointed to receiving virtues as gifts from the Holy Spirit that are bound up with an eschatological vision of the future.

It is such pneumatological enhancement that needs to become the true goal of humanity, for it is directed not so much towards self-centred desires, as towards the perfection of Christ in God. Wisdom, in particular, is about human relationship with God, and thus in some sense coheres with the emphasis on relationship that Jackelén argues is important when considering what happens after death. The virtue of wisdom can be distinguished from prudence, or practical wisdom, even though the two are closely interrelated.[42] Prudence is also relevant in as much as it is concerned with ethical decision-making where there are competing demands for attention. Aquinas recognized that the reality of sin meant that there are distortions in the possible attainment of virtues. Imperfect prudence would be to narrow the good to particular goods or individuals. Those arguing for enhancement technologies that benefit relatively few persons narrows the good attained, so that prudence is rendered imperfect. By contrast, rightly conceived prudence looks to individual, familial and political prudence, all aimed at the common good. Knowing which technologies are appropriate to develop or not is the task of prudence.[43] Yet prudence is also informed by the theological virtues. These virtues are *charity*, understood as fundamentally friendship with God, *faith* as fundamental trust in the God who is the creator of all that exists, and *hope* understood in terms of the ultimate as well as penultimate goals in human life and beyond.

As well as the virtue of prudence, the cardinal virtue of justice is worth mentioning in the context of secular drives for perfection. Forms of perfection that are detached from justice are incoherent from the point of view of Christian eschatology. The notion of God's righteous judgement is a prelude to the coming of the kingdom of peace as expressed in the Hebrew vision of the future. Justice is that habit of mind that renders each person his or her due, and interpreted as a virtue goes beyond simple rendition in terms of principles or rules, even if in practical terms justice is explicitly expressed through those rules. In this sense, justice understood as a *virtue* is more consistent with an eschatological vision that is more than simply just rewards and punishments, rather 'divine righteousness...is a righteousness that creates justice and puts people right, so it is a redemptive righteousness'.[44] The vision of the future includes the idea, in other words, of *Shalom*,

---

42   Celia Deane-Drummond, *The Ethics of Nature* (Oxford: Blackwell, 2004).
43   For a discussion in relation to biotechnology generally, see Deane-Drummond, *The Ethics of Nature*, pp. 86–110. For a discussion in relation to alternative genetic technologies, see C. Deane-Drummond, *Genetics and Christian Ethics* (Cambridge: Cambridge University Press, 2006).
44   Moltmann, *The Coming of God*, p. 335.

of the flourishing of God's creatures as they exist in relationship to each other and in relation to God. In other words, Christian eschatology is not just concerned about a better future, but also about redemption of the past and present, including injustices. It is about, as Moltmann has suggested, creating a just future, so that 'Christians anticipate the future of the new creation, the kingdom of justice and freedom, not because they are optimists, but because they trust in the faithfulness of God'.[45]

Both enhancements of human capabilities achievable through genetic technology and enhancement of human lifespan achieved through scientific research into combating ageing raise important ethical issues associated with justice. Peter Wenz has attacked the argument that genetic enhancement is compatible with a just and humane society.[46] Wenz also notes the motivation of the agents involved and is also correct in the need to distinguish between the drive for the kind of health-related enhancements that lead to protection against disease, and those that are desire related, that give people something that they desire other than health, such as height, intelligence or beauty. The latter could be compared with cosmetic surgery. He argues that if, for example, enhancements such as height are commonly available, then the result would be self-defeating as everyone will simply be taller. But in those cases where height enhancement is only available to richer people, this then widens further socioeconomic gaps. Health-related enhancements would also advantage some more than others, where such enhancements are only subject to market forces and the ability to pay. Other enhancements under discussion include the ability to resist diseases associated with exposure to pollutants such as PCBs. Yet the pollutant effects would be universal, so we can with Wenz infer that 'it is absurd to suppose that in the real world these people will be protected with genetic enhancements when at present our food-rich world allows 40,000 children to die every day from the effects of malnutrition'.[47] Such enhancements may be *legal* in the sense of being compliant with existing positive law, but they fail to express justice as *virtue*, as they undermine the wellbeing or flourishing of the community as a whole.

Anti-ageing technologies fit more easily in the category of enhancement of desire, rather than enhancement of health, though because longevity may well be associated with reduced quality of life, including chronic illnesses such as dementia, it is situated somewhat uneasily between these categories. No one has yet undertaken a full survey of public attitudes to these technologies, and such a survey would seem to be highly desirable.[48] John Harris's suggestion that people will want to exchange quality of life for longevity is simply unfounded.[49] Extending both the maximum and average human lifespan will have huge social,

---

45   J. Moltmann, *Creating a Just Future* (London: SCM Press, 1989), p. 8.

46   Peter Wenz, 'Engineering Genetic Injustice', *Bioethics* 19.1 (2005), pp. 1–11, www.blackwell-synergy.com/loi/biot.

47   Wenz, 'Engineering Genetic Injustice', p. 10.

48   Jayne C. Lucke and Wayne Hall, 'Who Wants to Live Forever?', *EMBO Reports* 6.2 (2005), pp. 98–102.

49   John Harris, 'Immortal Ethics', *Annals of New York Academy of Science*, 1019 (2004), pp. 527–34.

political and ethical consequences that arguably work against the common good, for it intensifies still further the strain on basic resources, especially in the Western world where there are demographic shifts towards an ever ageing population. The protagonists of this technology ignore such consequences and focus on short-term gain in terms of the individual persons concerned. If the damage caused by basic metabolic and environmental processes could be continuously repaired, some believe that basic maintenance of human individual function could continue for as much as several thousand years. Most resist the idea of immortality as such, though, as we noted above, such is the language used by transhumanists. There are those in established medical positions who are prepared to argue for significant gains in lifespan on the basis that ageing is undesirable, even if it is not a 'disease' as such.[50]

Remarkably, Aubrey de Grey, a biomedical gerontologist at the University of Cambridge, believes that anti-ageing technology is necessarily 'just' as in the Western world it will mean less expenditure on keeping people alive in the final years, freeing up funds for 'rejuvenation treatment'. This is a narrow interpretation of justice understood in terms of market factors and patient autonomy that is detached from wider, global considerations and the common good. He also considers that it would be entirely proper to conduct experimental anti-ageing techniques on people in their fifties or sixties who live in parts of the world where there are less stringent medical ethics procedures. He claims that:

> I don't have much time for the Hippocratic Oath myself. I think it's something that made a lot of sense when the understanding of medicine was primitive and people could spontaneously recover from illnesses for reasons that the doctor could not identity. That's where the 'do no harm' idea comes from. That becomes less reasonable as we become more knowledgeable about how to intervene in the body's metabolism. One also has to remember that around the world there are very different versions of medical ethics. There's good reason to believe that many of these therapies will first be developed in countries where they are more forward looking about cost-benefit ratio.[51]

In other words, Grey makes the remarkable suggestion that those who are less squeamish about ethical objections in other parts of the world will be used as experimental subjects in order to foster a technology that will ultimately benefit Western societies the most. Such a suggestion calls for widespread scrutiny and public debate, not least because of the issues of exploitation that it raises.

5. Conclusions

I have suggested in this chapter that enhancement technologies are not new in as much as they represent manifestations of a continual cultural drive towards perfection.[52] As such they betray an implicit secularized eschatology, one that is

---

50 EMBO, 'Interview: Curing Ageing and the Consequences', *EMBO Reports* 6.3 (2005), pp. 198–201.

51 EMBO, 'Interview; Curing Ageing and the Consequences', p. 200.

52 See also Maureen Junker-Kenny's contribution to this volume.

bent towards goals that cannot be attained merely within finite human existence. One might even view such trends as a fragmented search for immortality, but one that is detached from any organized 'life plan' as such. Furthermore, such trends are invasive in as much as prior commitments to healing particular diseases in medical contexts may be taken over by a more pervasive desire for enhancements. It is rare for medical ethicists to consider the implications of such developments, for, rather than focusing on wider social trends in medicine, they are more likely to concentrate on particular individual case studies, ensuring that proper procedures for consent have been met. The history of oestrogen use in the United States is a good example of how what was once a curative search for a particular medical condition soon became hijacked by promises of enhanced quality of life, defined in terms of unrealistic stereotypical ideas about attractiveness and beauty in women. The marketing potential for such drugs was too great for pharmaceutical companies to resist, many of whom also enticed less than scrupulous doctors to support the campaign. The slow realization that such drugs were only useful for a relatively small group of patients is testimony to the pervasiveness of cultural ideals of perfectability. Those technologies that promise enhancement against ageing itself deal more obviously with the question of immortality, though the ecological and environmental questions associated with such a trend are rarely discussed.

I have argued in this chapter that the underlying cultural issues associated with enhancement technologies need to be addressed. Christian theology can make a contribution to this discussion by outlining an eschatological perspective that is critical of secular alternatives. A fundamental hope for Christian believers is hope in the resurrection. Christianity is also realistic about the possibility of human sin, and the need to tackle attitudes that lead to the breakdown of communities through all forms of injustice and violence. I have indicated that the classic tradition of deification and participation in God is relevant here, for it permits the goal of perfection to be expressed, but takes it up into a theological category that moves the self away from itself towards God. Such a doctrine can be criticized on the basis that it denies individual expression, or that it leads to detachment from worldly concerns. I have resisted both interpretations. Rather, I suggest that the aspiration for theosis provides the incentive to go beyond virtues as simply learned in the human community, and recognize virtues as also attainable through the gift of God. In particular, virtues such as prudence, charity and justice are particularly relevant in the midst of voices that would claim passive acceptance of the status quo regarding conditions of injustice.

Chapter 13

SAVING US FROM OURSELVES: CHRISTOLOGY, ANTHROPOLOGY
AND THE SEDUCTION OF POSTHUMAN MEDICINE

Brent Waters

Contemporary medicine is making what may be characterized as a posthuman
move. This move can be seen in the advent of regenerative medicine, especially
in respect to its potential for enhancing physical and cognitive capabilities.
The underlying rationale for such enhancement shares much in common with
posthuman discourse, which presumes that humans are highly malleable beings
who are not defined or delineated by any given boundaries or limits. Anticipated
advances in regenerative medicine, therefore, offers an opportunity for humans to
transform themselves into superior beings. This chapter both examines this relationship between regenerative medicine and posthuman discourse, and critiques
it from a Christian theological perspective.

1. *Posthuman Medicine*

Proponents of regenerative medicine foresee an age of greatly expanded longevity.
Moreover, humans will not only enjoy extended lifespans, but will also maintain
their physical strength and mental abilities. We should expect our descendants
to live longer without the burdens associated with growing older. These benefits
will be achieved through a combination of more efficacious drugs, cellular and
tissue rejuvenation resulting from stem-cell research, and genetic manipulation.
Although the possibility of infinite rejuvenation cannot be ruled out in principle,
human biology may not prove to be as malleable as assumed by the more enthu-
siastic champions of regenerative medicine. If this proves to be the case, then
biotechnology can be combined with other technologies in order to move beyond
natural limits.

A number of these applications have already been developed. Sophisticated
prosthetic arms and legs have, in some cases, restored up to 90 per cent recovery
of dexterity and mobility. Nanotechnology is being used to construct better stents
and artificial blood veins, reducing the need for more invasive surgery in treating
heart disease. An electronic device implanted in the brain enables quadriplegics to
move the cursor and click icons on a computer screen. With further technological
development each of these therapeutic techniques could lend themselves to future
enhancement applications. Prosthetics could improve the strength, mobility and
dexterity of athletes or soldiers. Heart disease could be prevented, and more efficient cardiovascular systems might be constructed, by using nanotechnology to
construct more durable organs and arteries. Neuro-links between the brain and

computers could improve interactive speed, and enhance the experience of virtual reality programs.

One effect of these therapeutic applications will be modest gains in longevity for their beneficiaries as a result of improved healthcare. Anticipated enhancement applications are admittedly speculative, but if their potential is realized then greatly expanded lifespans might prove feasible. Indeed, ageing itself is being perceived as a 'disease' that regenerative medicine is designed to treat. If ageing is a disease, however, why can't it be 'cured' rather than merely treated? Again, some of the more vociferous champions of regenerative medicine and biotechnology see no reason why such a cure cannot be found.[1] Yet if ageing can, at least in principle, be cured, does this not imply that the advent of regenerative medicine is tantamount to declaring a war against death? Presumably the answer must be 'yes', for the end of bodily degeneration and morbidity is mortality. But what would victory mean, and what would be the cost? Total victory would result in immortality, and if this ambitious goal proves illusive, greatly expanded longevity would be a partial, but nonetheless significant, triumph. The cost would entail the radical transformation of medical practice, *and* the patients it transforms. Waging a total war against death requires medicine to forsake its traditional emphases on care, treatment and prevention in favour of enhancement. Medicine would no longer focus on preventing illness and then treating and caring for patients suffering the inevitable ravages of deteriorating bodies, but to eliminate the organic sources of their distress. The role of medicine would not be to assist patients to come to terms with their mortality, but to enable them to vanquish death, or at least keep it at bay for as long as possible. Moreover, if an effective campaign against death is to be waged, then medicine must also transform its patients, for it will be forced to forsake its commitment to relieving the human condition in order to radically alter it.

If regenerative medicine may be characterized as a first step in curing ageing, and thereby preventing or delaying death, then a provocative issue is forced upon us: should humans use their technology to become better than human? It would seem that some such aspiration is at play if the goal is to use technology to overcome or extend the finite and mortal limits that evolution has programmed into human biology. If these limits are overcome or greatly extended, however, then mortality is effectively removed as a definitive feature of human life. Yet in the absence of this feature, what are humans aspiring to become as artefacts of their own engineering? Or to pose the same question more starkly: should medicine help us become posthuman?[2]

---

1   The perception of ageing as a disease is not prevalent among medical practitioners, but is more often invoked by ardent proponents of biotechnology or digital technologies. These proponents perceive early developments in regenerative medicine as the first, cautious step in a much more ambitious enterprise that will transform standard medical care over time. See Stephen S. Hall, *Merchants of Immortality: Chasing the Dream of Human Life Extension* (Boston, MA and New York: Houghton Mifflin Co., 2002), and Ray Kurzweil, *Singularity is Near: When Humans Transcend Biology* (New York: Viking, 2005).

2   I use the term 'posthuman' to refer broadly to a goal of improving human longevity and performance through the application of various technologies. Although transhumanists share many

The prospect of such radical transformation has not received a universally warm reception. Leon Kass, for example, sees a posthuman future as an assault against human dignity. Natural necessity should be struggled with and met, not eliminated. Without natural limits and constraints, individuals are stripped of their dignity. 'Human aspiration depends absolutely on our being creatures of need and finitude, and hence longings and attachments.'[3] Becoming posthuman would require humans to abolish their humanity. Indeed, no true 'friend of humanity cheers for a posthuman future'.[4]

Francis Fukuyama is afraid that the extensive interventions being envisioned by the more ardent proponents of regenerative medicine may undermine the biological foundation of liberal democracy. Enhancing the performance of the body and mind means that human nature is also being transformed, a perilous endeavour that should be resisted. The gist of his argument can be captured by summarizing two substantive claims. First, any meaningful discourse on human rights must be grounded in human nature, which he defines as the 'sum total of the behavior and characteristics that are typical of the human species, arising from genetic rather than environmental factors'.[5] Individuals, civil societies and political communities are not created *ex nihilo*, but are derived from innate behavioural characteristics. A natural moral sense has evolved over time as demonstrated in a range of emotive responses that is 'species-typical'. Most human adults, for instance, respond protectively to imperilled children, and are repulsed by acts of child abuse.

Second, dignity is not an abstract concept, but a natural quality bequeathed by a genetic endowment that is uniquely human. It is also an endowment that promotes emergent, rather than reductive, forms of behaviour among individuals and groups, and any attempt to separate the parts from the whole would result in disfiguring a human nature that has been selected by natural evolution. Altering genes, albeit for therapeutic reasons, is nonetheless also altering human nature. Tinkering with this uniquely human genetic endowment could negate the civil and political rights of liberal democracies that natural selection has made possible. Consequently, any prospect of a posthuman future should be resisted, because 'we want to protect the full range of our complex, evolved natures against attempts at self-modification. We do not want to disrupt either the unity or the continuity of human nature, and thereby the human rights that are based on it.'[6]

Jürgen Habermas objects to the prospect of selecting or enhancing the genes of offspring, because it violates the autonomy of the purported beneficiaries. Genetically enhanced children cannot give their free and informed consent to

---

of the goals of this posthuman agenda, they incorporate ideological commitments not widely shared by other proponents. My comments are directed toward a more general posthuman vision rather than a more specific transhumanist programme.

3   Leon R. Kass, *Life, Liberty and the Defense of Dignity: The Challenge for Bioethics* (San Francisco, CA: Encounter Books, 2002), p. 18.

4   Kass, *Life, Liberty and the Defense of Dignity*, p. 6.

5   Francis Fukuyama, *Our Posthuman Future: Consequences of the Biotechnology Revolution* (New York: Farrar, Strauss and Giroux, 2002), p. 14.

6   Fukuyama, *Our Posthuman Future*, p. 173.

selections and alterations that are conducted at the embryonic stage of their development. Consequently, every person should have the right to be born with an unaltered genome. The power that parents could wield over their children by using such transformative technology is tantamount to tyranny, thereby undermining the 'normative and natural foundations' of civil society in favour of indulging the narcissistic desires of parents wanting perfect children.[7] Habermas believes this issue so urgent that it cannot be left to 'biologists and engineers intoxicated by science fiction'[8] or 'self-styled Nietzscheans'[9] masquerading as posthumanists.

Do these objections provide sufficiently good reasons to prevent regenerative medicine from initiating a war against ageing and mortality?

2. Posthuman Discourse

Appeals to dignity, human nature and rights no longer have sufficient moral weight in late modernity to dissuade humans from transforming themselves into posthumans. Invoking such essentialist and foundational principles is to erect a bulwark that has already been dismantled by postmodern critique.[10] There are no given features, such as finitude and mortality, which define the quality and character of human life and lives. Personal, social and political identities are subjected to continuous deconstruction and reconstruction. In this respect, medicine can be used to deconstruct and reconstruct human bodies, and by extension social and political bodies as well. Consequently, the objections raised by Kass, Fukuyama and Habermas can be simply rendered irrelevant by posthumanists rather than engaged directly.

According to Elaine Graham, for instance, we have not yet committed ourselves to becoming posthuman, but we are starting to entertain the possibility. She argues her case by examining recent developments in representative sciences, technologies and literary genres. Cybernetics, for instance, provides a scientific basis regarding the malleability of human minds and bodies. What constitutes human life, both individually and corporately, can be reduced to patterns of information that, with the aid of ever more sophisticated technology, can be manipulated into virtually limitless configurations. The technological application of this cybernetic principle is not confined to the construction of virtual realities that enhance the mind, but through a combination of biotechnology and sophisticated prosthetics that can be used to reconstruct the body. The posthuman is a cyborg in which the line separating the organic and the mechanistic is eliminated. Graham contends that this posthuman vision is a cybernetic and libertarian version of an earlier social Darwinism, a theme often rehearsed in modern science-fiction literature celebrating an 'obsessive attempt to cheat death and to place humanity in a posi-

---

7   Jürgen Habermas, *The Future of Human Nature* (Cambridge: Polity Press, 2003), p. 20.

8   Habermas, *The Future of Human Nature*, p. 15.

9   Habermas, *The Future of Human Nature*, p. 22.

10   See Richard J. Bernstein, *The New Constellation: The Ethical-Political Horizons of Modernity/Postmodernity* (Cambridge, MA: MIT Press, 1992).

tion of mastery and dominion over non-human nature'.[11] Consequently, there is little certainty whether the emerging posthuman will be perceived as a superior being to be celebrated, a monster to be feared, or perhaps both. Graham concludes that the posthuman is becoming an increasingly important symbol, if not cultural icon, that 'confounds but also holds up to scrutiny the terms on which the quintessentially human will be conceived'.[12]

N. Katherine Hayles insists that we are already committed to this transformative enterprise, because we now think of ourselves as posthumans. We live in a cybernetic world of infinitely pliable information, and therefore a world of infinite possibilities to be explored and actualized. There is no absolute boundary separating 'bodily existence and computer simulation, cybernetic mechanism and biological organism, robot technology and human goals'.[13] Evolution has given humans a body as a severely limited and inefficient prosthesis of the will, a situation that can now be corrected through technological development. We have reached a point where the human evolutionary process, in both its biological and cultural forms, can, and should, be guided through wilful rather than natural selection. Consequently, there is little question that medicine can, and should, be used to construct posthuman bodies and identities. Her only plea is that we retain some semblance of embodied finitude, though she does not appeal to any normative reasons for doing so. Rather, she declares that being human means being embodied, and that the symbiotic relationship between humans and machines will prove to be limited. Despite these reservations, Hayles is optimistic that a posthuman future will not precipitate the end of humanity. 'What is lethal is *not* the posthuman as such but the grafting of the posthuman onto a liberal humanist view of the self.'[14] If this deadly combination can be avoided, a posthuman future need not be feared, for it depends on how it is constructed, and is therefore subject to our control.

Donna Haraway has no reservations in championing a posthuman world, for it heralds a new era of liberation, especially for women. The concept of 'nature' has not been kind to women, and reconstructing themselves as cyborgs will free them from their 'natural' captivity. In becoming cyborgs all so-called natural boundaries and distinctions are breached and nullified, disclosing a new social and political reality 'in which people are not afraid of their joint kinship with animals and machines'.[15] Such kinship frees individuals from the biological and cultural constraints preventing one's self-creation, for the biological and cultural are collapsed into a singular, constructed reality. 'The cyborg is a kind of disassembled and reassembled, postmodern collective and personal self.'[16] A posthuman future

---

11 Elaine L. Graham, *Representations of the Post/Human: Monsters, Aliens and Others in Popular Culture* (Manchester: Manchester University Press, 2002), p. 64.

12 Graham, *Representations of the Post/Human*, p. 11.

13 N. Katherine Hayles, *How We Became Posthuman: Virtual Bodies in Cybernetics, Literature, and Informatics* (Chicago and London: University of Chicago Press, 1999), p. 3.

14 Hayles, *How We Became Posthuman*, pp. 286–87, emphasis added.

15 Donna J. Haraway, *Simians, Cyborgs, and Women: The Reinvention of Nature* (New York: Routledge, 1991), p. 154.

16 Haraway, *Simians, Cyborgs, and Women*, p. 163.

effectively has no origin or determined future, no need for community or interest in remembering, thereby eradicating any lingering notion of 'salvation history' in either its sacred or profane guise.[17] Consequently, regenerative medicine is welcomed as the first step in supplying the technological tool chest with what will be needed in constructing posthuman bodies and identities.

Given these emphases on fluidity and construction, Kass is hard-pressed to make a convincing case against posthuman transformation. Although he champions the finite and mortal nature of human life, he is equally dedicated to partaking the advances of modern medicine. He has no objection to the prospect of living a long and healthy life, surrounded by loving children and grandchildren. To a large extent, medicine helps us achieve this good goal through the development of better preventive and therapeutic techniques. Improved healthcare is no enemy of human dignity. Yet this admission blurs the lines separating prevention, therapy and enhancement. Reducing cholesterol, for example, helps prevent a heart attack. Bypass surgery is a therapy that helps restore the function of a damaged cardio-vascular system. Both procedures may also be regarded as enhancements, however, by presumably extending the length and improving the quality of the patient's life. Kass cannot resolve a fundamental dilemma which is entailed in his twin commitments to improving healthcare while also preserving finitude and mortality, for the improved prevention and therapy that he approves pushes back the finite and mortal limits that he fears. The only way to resolve this dilemma would require a curious restraint or withholding of legitimate therapeutic applications. Suppose, for example, that artificial arteries and heart valves were developed that when surgically implanted in an 80 year old suffering heart disease would restore cardio-vascular functions to roughly that of a healthy 40 year old. To honour Kass's commitment to finitude and mortality, should we not insist that the artificial devices be designed to only restore the cardio-vascular functions to that of a healthy 80 year old? The patient would quite rightfully object that a therapy is being withheld solely on the basis of age, violating the standard medical principle of non-discrimination. Kass wants to draw a line separating enhancement from prevention and therapy, but he cannot do so on a plane constantly shifting in response to technological development. The only way to stop this movement would be to set an outside limit for longevity toward which medicine should aspire but go no further in developing and applying preventive and therapeutic treatments. But Kass is either unable or unwilling to state this limit. Consequently, his objection to the prospect of posthuman transformation appears to be more an emotive assertion rather than any normative appeal to finitude and mortality being constitutive elements of a so-called dignity he wishes to protect.

Fukuyama's concern is not to defend an innate human dignity, but to preserve the underlying biological elements upon which liberal democracies are based and from which dignity is derived, and in which it is enacted. The threat of posthuman transformation is that it will destroy the evolutionary foundation, thereby threatening the political and social derivatives. His argument, however, incorporates a paradox. If liberal democracy is predicated on evolutionary development at a

---

17   See Haraway, *Simians, Cyborgs, and Women*, pp. 157–58.

particular point in time, and if posthuman transformation should be prohibited on the grounds that it would disturb this development, then does this not imply that natural selection should also be controlled in order to prevent the possibility of an equally threatening natural mutation? If the answer is 'yes', then Fukuyama is in the rather awkward position of contending that human evolutionary development should proceed to a certain point, but no further. The feasibility of imposing such a moratorium is not only highly unlikely, but it denies the very genius of the natural process that he is purportedly defending. If the answer is 'no', then he is hard-pressed to demonstrate why natural selection is necessarily superior to wilful transformation. Human biological and cultural evolution may, through natural selection, reach a stage where liberal democracies are neither needed nor beneficial, so the prospect of posthuman transformation cannot be condemned solely on the basis that it might imperil a particular social and political order. If Fukuyama's concern is that the pace of posthuman transformation would prove socially and politically destructive, then he is pleading for regulation rather than prohibition. But such a plea cannot rule out in principle the use of medicine to enhance performance, or even transform humans into superior beings, albeit at a more deliberate pace.

Habermas's objection to posthuman transformation is one of procedural justice: altering the genes of other persons without their consent violates their autonomy. Consequently, every person has the right to be born with an unaltered genome. This principle has little moral purchase, however, since Habermas has already conceded that based on the same principle of autonomy, women have the right to pursue assisted reproduction, and abort 'defective' foetuses. In both instances, certain genetic traits are being selected for or against, a process which Habermas seemingly finds distressing in regard to the prospect of genetically enhancing offspring. He has already granted the chief tenet of so-called procreative liberty that every person has the right to pursue her reproductive interests, and that technological intervention is often needed to exercise this right.[18] Yet, in implicitly accepting this premise, he also effectively undermines his objection: if a person has the right to have a child, then she also has the right to have a desirable child. Exercising this right necessarily entails selecting for and against certain traits, and there is no compelling reason why the desirable traits that have been selected should not be enhanced. Placing constraints on enhancing offspring would only serve to deny the rights of individuals to pursue their reproductive interests, thereby restricting their autonomy. The principle of autonomy that Habermas invokes as the centrepiece of his objection is used to negate his subsequent argument. His complaint against the prospect of enhancing offspring is effectively reduced to an attempt at grafting a liberal anthropology on to the emerging posthuman, precisely the kind of move that Hayles fears will prove lethal.

If appeals to dignity, democracy and autonomy cannot support convincing objections to posthuman transformation, can a more engaging theological objection be crafted?

---

18   See John A. Robertson, *Children of Choice: Freedom and the New Reproductive Technologies* (Princeton, NJ: Princeton University Press, 1994).

3. *Theological Discourse*

There are two reasons why it should not be assumed that Christian theology is *necessarily* opposed to the prospect of posthuman transformation. First, there is nothing in the Christian tradition that is predisposed toward defending an innate human dignity, liberal democracy, or autonomy, at least as propounded, respectively, by Kass, Fukuyama and Habermas. Human dignity is not an inherent quality, but is derived from the gift of grace given by God in Christ. Freedom is not bestowed by a political community, but is discovered and embraced through obedience to Christ, and is embodied in varying forms of government that are appropriate to particular cultures and historical circumstances. Autonomy is not the absence of external constraints upon one's will, but is the recognition of limits and boundaries defining one's role and participation in various associations, especially the Church as the body of Christ.

Second, posthumanists and Christians share a number of common interests and concerns. Both agree, for instance, that the current state of the human condition is far from ideal. For posthumanists, humans have not achieved their full potential, while for Christians humans have not yet become the kind of creatures God intends them to be. In response, both agree that humans need to be released from their current plight. For posthumanists, this is achieved through technologically driven transformation, while Christians believe they are transformed by their life in Christ. Both agree that death is the final enemy. One conquers this foe by extending longevity and perhaps achieving virtual immortality,[19] while the other is resurrected into the eternal life of God. Consequently, both place their hope in a future that at present appears as little more than a puzzling reflection in a mirror; one can only speculate what life will be like in a posthuman world, or a new heaven and earth.

We may say, then, that posthumanists and Christians share, in broad outline, a metanarrative that is transformative in character, but they differ sharply over the unfolding storyline. How may we account for both this similarity and difference? The underlying soteriological and eschatological presuppositions of posthumanism are drawn from an eccentric combination of attenuated Platonic and materialistic principles.[20] Posthumanists share with Plato the belief that the soul (now read mind) is imprisoned within the confines of a finite and mortal body. The realization that death is the only escape leads to despairing about one's embodiment. This despair is overcome through philosophic contemplation that infuses life with meaning in preparing oneself for an inevitable death that liberates the soul. It is at this point that posthumanists part company with Platonists. As materialists they do not believe there is any soul that can survive the death of the body. The

---

19   Virtual immortality can purportedly be achieved by uploading an individual's mind or consciousness, and then downloading into a robotic body or other substrata. See Ray Kurzweil, *The Age of Spiritual Machines: When Computers Exceed Human Intelligence* (New York and London: Penguin Books, 2000), and Hans Moravec, *Mind Children: The Future of Robot and Human Intelligence* (Cambridge, MA and London: Harvard University Press, 1988), and *idem, Robot: Mere Machine to Transcendent Mind* (Oxford and New York: Oxford University Press, 1999).

20   See, e.g., Moravec's appeal to Platonic themes in *Robot*, pp. 196–98.

mind – as the source of an enduring identity – is unfortunately dependent upon the brain, and with its death a person ceases to exist. This constraint, however, can be relaxed, if not eventually overcome, by human ingenuity; a more effective war against decay and death can be waged. Since the body consists of underlying information, cybernetic technology can be developed and utilized to create more enduring patterns over time. The body can be transformed into a more durable, if not virtually immortal, prosthetic for the mind. In short, posthumanists place their hope in transforming flesh into data.

In contrast, the Christian narrative claims that the eternal has entered and redeemed the temporal; the Word became flesh. Through the Incarnation, God has redeemed creation from its futility, thereby negating death as witnessed by the resurrection of Jesus Christ from the dead. It is the empty tomb that differentiates Christian hope from its posthuman counterpart. The task is not to rescue the soul (or mind) from the body. Rather it is as embodied creatures that in Christ we are redeemed and participate in the eternal life of the triune God. The doctrine of Christ's bodily resurrection, therefore, should not be casually discarded as a relic of a credulous age, for it serves as a reminder that the body is God's good gift, and not a shell to be despised. For Christians, death is a real enemy but it is not a source of despair, for in Christ death has already been overcome within eternity. What separates Christian from posthumanist hope is that the latter seeks immortality, while the former awaits eternal life.[21] Transformation does not consist of greatly extended longevity, perhaps culminating in virtual immortality, but in a temporality, and its accompanying finitude and mortality, that has already been transcended by eternity. In this respect, it is telling that Hayles wants to affirm and preserve finitude and embodiment in constructing a posthuman future, but she is either unable or unwilling to invoke any normative reasons for doing so. Her only appeal is that such affirmation and preservation will enhance the prospects for survival, but surviving for what end or purpose is never mentioned. Her celebration of the finite body is more a wake for the death of the human than a party greeting the arrival of the posthuman that she cannot even recognize. Christians in contrast wait for a complete fellowship with a living Lord whom they encounter, albeit partially, in the sacraments and life together in the Spirit. Finitude and embodiment are genuinely celebrated because it is through these means that the temporal and the eternal are mediated in the Incarnation.

These differing perceptions of finitude and mortality, particularly in respect to the body, reflect strikingly dissimilar anthropologies presupposed by posthumanists and Christians. For posthumanists, *homo sapiens* is a crude, transitional species. As Nick Bostrom asserts, we should 'view human nature as a work-in-progress, a half-baked beginning that we can learn to remold in desirable ways'.[22] This remoulding is the practical enactment of the imperative of self-improvement. Such non-invasive measures as improved diet and healthcare are modest steps in

---

21   For an analysis of the difference between eternity and immortality, see Hannah Arendt, *The Human Condition*, introduction Margaret Canovan (Chicago and London: University of Chicago Press, 2nd edn, 1998), pp. 17–21.

22   Nick Bostrom, 'Transhumanist Values', www.nickbostrom.com/ethics/values.html.

the right direction, but they fail to maximize an individual's full potential. The human species is perhaps becoming a bit more baked, but still requires more cooking. The problem is that, given biological limitations, humans do not have enough time to be sufficiently cooked, thereby restricting their improvement as both individuals and a species. Short lifespans must be greatly extended if humans are to emerge as the superior posthumans they have the latent potential to become. Consequently, regenerative medicine is to be warmly embraced as an initial step toward overcoming the twin threats of finitude and mortality.

The immediate culprit to be vanquished is human biology. The human body, in its present state, is too fragile a vessel to allow the full development of the mind's potential. A more robust host must be constructed that will provide the necessary time to enable this development. Although posthumanists reject a religious notion of a fall, they nonetheless offer a salvific story. Through natural selection evolution has formed a species in which the mind is the most distinguishing trait. Yet humans remain hampered by severe biological limitations as witnessed by their finitude and mortality. If the mind is to flourish, humans must seize control of their future evolution by transforming themselves into superior beings. If humans are to rescue themselves from their finite and mortal limits, they must clutch the salvation that technological development places in their grasp.

We may characterize this posthumanist anthropology as *paradoxically materialist and dualist*. In one respect, posthumanists are thoroughgoing materialists in that they are confident that the body can be engineered to endure for a greatly extended period of time. But they are also dualists in that they believe that a life of the mind exists, to a large extent, independently of, or despite, the material limitations of the body in which it is encased. Paradoxically, the materialism posthumanists purportedly affirm is simultaneously also the object of their despair. The so-called 'prosthetic' of the mind is both its host and jailer. Moreover, no amount of technological transformation can completely resolve the dilemma, for all material constructs remain subject to the limits of finitude and mortality. These constraints can, perhaps, be greatly relieved, but they cannot be eliminated short of negating the material itself. In endeavouring to liberate the mind from its finite and mortal captivity, posthumanists also undermine the material substance which sustains it.[23] Posthumanists are trapped by an anthropology that is simultaneously, and curiously, Pelagian and Manichean. These tendencies are apparent in Haraway's cyborg, for her construct represents the triumph of the will. In this artefact the finite and temporal limitations upon the will are overcome by eliminating the boundary separating nature from artifice. Yet the erasure itself necessitates the negation of the construction materials. The will remains constrained by technological potential, and cannot be perfected until liberated from this constraint. The logical conclusion of Haraway's construction project is a virtual world in which

---

23 Kurzweil's and Moravec's belief that the mind can exist in substrata other than a particular body seemingly overcomes this dilemma. This is a rather dubious belief, however, given recent developments in the neurosciences. See, e.g., Malcolm Jeeves (ed.), *From Cells to Souls – and Beyond: Changing Portraits of Human Nature* (Grand Rapids, MI and Cambridge: Eerdmans, 2004).

the will is at last perfected in its freedom from all finite and temporal limitations. Consequently, the cyborg is, like *homo sapiens*, a transitional species that is being drawn toward its redemption from the body. In this respect, Hayles is effectively rendered an enemy rather than an ally, for her finite and embodied posthuman shares more in common with Kass's and Fukuyama's biological essentialism than with the disembodied and infinite future that Haraway's cyborg anticipates.

This anthropological captivity results in a conflation of distinguishable concerns that distorts an otherwise good desire. There is nothing inherently wrong in wanting to live a long and healthy life. But when finitude and mortality are perceived as hostile assailants against, rather than definitive features for, realizing this good objective, the healthy desire mutates into a cancerous concupiscence that ultimately destroys both the object of desire and the one desiring it. Pursuing a long and healthy life does not entitle one to regard ageing as a disease to be cured, or death as a proper target of medical warfare. Taken to its logical conclusion, such assumptions lead to the inevitable conclusion that embodiment itself is a disabling handicap. In short, humans cannot save themselves from being the creatures that are by definition temporal.

In contrast, Christian anthropology may be characterized as *redemptively materialist and non-dualist*. As creatures, humans are material beings, subject to all the finite and mortal limits this status entails. Finitude and mortality are definitive features of this status, for in their absence humans would not be creatures. Consequently, to wage a war against ageing and death is to attack the very qualities that define what it means to be a creature, and therefore a denial of one's being. But is this not simply a rhetorical move to reassert Kass's claim shorn of its dignity that is purportedly rejected by Christians? No, for it is a Christological claim rather than an assertion about human nature. In the Incarnation, the Word is made flesh, an act of divine initiative that simultaneously affirms and redeems the created order. In Emmanuel – God with us – the creating Word reaffirms what it has created as good, a declaration reinforced by the vindication of created order in the resurrection of Jesus Christ from the dead as the culmination of the Incarnation.[24] In this respect, finitude and mortality are not unfortunate limits that constrain human flourishing, but essential elements sustaining creation and its creatures over time. The act is also redemptive in that the affirmation and vindication are eschatologically oriented, pointing toward creation's destiny. Since creation is temporal it must and will eventually cease to exist, but it is saved from futility by being resurrected into the eternal life of the triune God. Finitude and mortality are not, therefore, evil forces from which humans must be saved, but are characteristics of a creation that is being redeemed by its creator and saviour. In this restricted sense, death as a necessary consequence of temporality remains the final enemy, but it is an enemy that has already been conquered by the dwelling of the eternal with the temporal. This is why Christians place (or should place) their hope in the resurrection of the body instead of the immortal soul, for the former represents the completion of creation while the latter negates it. And it is with such

---

24   See Oliver O'Donovan, *Resurrection and Moral Order: An Outline for Evangelical Ethics* (Leicester: InterVarsity Press, 1986), pp. 31–52.

hope that humans may rest content with remaining creaturely rather than aspiring to transform themselves into something they were not created to be.

The implications of these contrasting anthropologies can be drawn out by focusing on the posthumanist fixation on mortality at the expense of natality. Following Hannah Arendt, birth and death are the two general conditions that both demarcate human existence and bind humans together over time.[25] It is the former, rather than the latter, which should provide the principal metaphor ordering human life. Natality promotes generational continuity while also encapsulating the possibility for change and improvement. Each new birth embodies both a continuous line of memory and a trajectory of anticipation; a self-giving which creates a recipient who is both like and yet unlike the giver. The gift of every parent is also the unique possibility of each child. Although death is not to be embraced warmly, mortality is not humankind's great curse. When death is perceived as nothing more than a malicious threat, natality is robbed of its power to renew and to regenerate. To be fixated on denying mortality is to promote an ordering of life that attempts to cheat fate for as long as possible. Survival then becomes a consuming and distorted desire that denigrates and corrupts any other value or consideration. The birth of a child holds no hope or promise, but serves only as a reminder of a cruel fate to be despised and despaired. Natality is simply an ugly confirmation of finitude and mortality. Consequently, personal survival, as opposed to biological and social reproduction, becomes the tyranny of the present generation exercised over subsequent ones.

It is precisely this posthuman tyranny inspired by its survivalist anthropology that Christian theology must resist, for it destroys an intergenerational fellowship that is consummated in eternity and not through the transformation of the human species. It is this posthuman seduction that medicine must also resist, for its purpose is not to save humans from their finitude, but to strengthen these bonds of fellowship by helping its patients come to terms with, but not overcome, their mortality. Again, this does not imply that there is anything necessarily wrong with desiring a long and healthy life which medicine properly assists us in satisfying. But such a desire when properly ordered does not entail a consuming fixation on surviving for as long as possible in which the natural process of ageing is distorted into a disease to be treated and cured. Rather, it is in our ageing that we are enabled to contribute something of ourselves to future generations, and if our longevity is extended it is an opportunity to be more, not less, generous in our giving. If medicine is to be truly *regenerative*, then its purpose is not to transform its patients, but to help them see that in their finitude and mortality lie the avenues for a genuine renewal of temporal life as it is drawn toward its destiny in the eternal life of God.

Does the preceding argument imply that Christians should not resist ageing and death? No, for in affirming the finite and embodied character of our status as creatures, medicine is a useful instrument in assisting us in pursuing our respective callings and vocations. Extended longevity and enhanced performance, however, are not the proper ends of medicine, but residual benefits. The proper goal of

---

25   See Arendt, *The Human Condition*, pp. 7–11.

medicine is not to improve the prospects of individual survival, but is a concrete act enabling the love of God and neighbour. Christians need not, then, specify that longevity should be extended to a certain age but no further, or that performance may be enhanced to this point but not beyond. For the principal issue at stake is not self-improvement, but how the self is enabled to serve the needs of neighbours. Consequently, anticipated advances and applications of regenerative medicine should be evaluated in respect to the extent that they either strengthen or weaken the bonds among neighbours. And the task of strengthening these bonds does not so much entail the enterprise of transforming ourselves and our neighbours as preparing creation and its creatures for their transformation in Christ.

BIBLIOGRAPHY

Agamben, G., *Homo Sacer: Sovereign Power and Bare Life* (Stanford, CA: Stanford University Press, 1998).

Alexander, B., *Rapture: How Biotech Became the New Religion* (New York: Basic Books, 2003).

Anderson, W.F., 'Genetics and Human Malleability', *Hastings Center Report*, 20 (February 1990).

—, 'Human Gene Therapy: Scientific and Ethical Considerations', *The Journal of Medicine and Philosophy* 10 (1985), pp. 275–91.

Anon., *DNDi (Drugs for Neglected Diseases Initiative): An Innovative Solution* (Working Draft), www.accessmed-msf.org/upload/ReportsandPublications/19 220031120226/DNDi.pdf (accessed 22 September 2005).

Aquinas, Thomas, *Summa Theologiae* (ed. and trans. Fathers of the English Dominican Provinces; 60 vols.; London: Eyre and Spottiswoode, 1964–76).

—, *Summa Theologiae*, XIV (London: Blackfriars in association with Eyre and Spottiswoode, 1963).

Arendt, Hannah, *The Human Condition*, introduction Margaret Canovan (Chicago and London: University of Chicago Press, 2nd edn, 1998).

Aron, A., *et al.*, 'Reward, Motivation, and Emotion Systems Associated with Early-Stage Intense Romantic Love', *Journal of Neurophysiology* 94 (2005), pp. 327–37.

Athenagoras, *A Plea for the Christians*, trans. B.P. Pratten, *Anti-Nicene Fathers Vol. II*, www.ccel.org/fathers2/ANF-02/anf02-46.htm.

Atkins, Margaret, 'For Gain, for Curiosity or for Edification: Why Do We Teach and Learn?', *Studies in Christian Ethics* 17.1 (2004), pp. 104–17.

Atwood, M., *Oryx and Crake* (London: Bloomsbury, 2003).

Badham, P., and L. Badham, *Immortality or Extinction?* (London: Macmillan, 1982).

Badmington, Neil, *Alien Chic: Posthumanism and the Other Within* (London: Routledge, 2004).

Banner, M., 'The Practice of Abortion: A Critique', in Michael Banner, *Christian Ethics and Contemporary Moral Problems* (Cambridge: Cambridge University Press, 1999), pp. 86–135.

Barbour, Ian G., *Nature, Human Nature, and God* (London: SPCK, 2002).

—, 'Neuroscience, Artificial Intelligence, and Human Nature', in Robert John Russell, Nancey Murphy, Theo C. Meyering, and Michael A. Arbib (eds.), *Neuroscience and the Person: Scientific Perspectives on Divine Action*, (Vatican City State and Berkeley, CA: Vatican Observatory and Center for Theology and Natural Sciences, 1999), pp. 249–80.

Barshinger, C.E., L.E. LaRowe and A. Tapia, 'The Gospel according to Prozac:

Can a Pill Do What the Holy Spirit Could Not?' *Christianity Today* (14 August 1995), pp. 34–37.

Barth, Karl, *Church Dogmatics*, III/3 (London: Centenary Press, 1960).

Bartos, Emil, *Deification in Eastern Orthodox Theology: An Evaluation and Critique of the Theology of Dumitru Stăniloae* (Carlisle: Paternoster, 1999).

Batement Novaes, S., and T. Salem, 'Embedding the Embryo', in J. Harris and S. Holm (eds.), *The Future of Human Reproduction: Ethics, Choice, and Regulation* (Oxford: Clarendon Press, 1998), pp. 100–126.

Bauckham, R. (ed.), *God Will be All in All: An Eschatology of Jürgen Moltmann* (Edinburgh: T & T Clark, 1999).

Bauman, Zygmunt, *Mortality, Immortality and Other Life Strategies* (Cambridge: Polity Press, 1992).

Beauchamp, Tom L., and James F. Childress, *Principles of Biomedical Ethics* (Oxford: Oxford University Press, 5th edn, 2001).

Beck, U., *Risk Society: Towards a New Modernity* (London: Sage, 1992).

Begbie, Jeremy S., *Theology, Music and Time* (Cambridge: Cambridge University Press, 2000).

Benedikt, Michael (ed.), *Cyberspace: First Steps* (Cambridge, MA: MIT Press, 1992).

Benhabib, Seyla, *Situating the Self: Gender, Community and Postmodernism in Contemporary Ethics* (Cambridge: Polity Press, 1992).

Bernstein, Richard J., *The New Constellation: The Ethical-Political Horizons of Modernity/Postmodernity* (Cambridge, MA: MIT Press, 1992).

Björklund, A., *et al.*, 'Neural Transplantation for the Treatment of Parkinson's Disease', *Lancet Neurology* 2 (2003), pp. 437–45.

Blanke, O., T. Landis, L. Spinelli and M. Seeck, 'Out-of-body Experience and Autoscopy of Neurological Origin', *Brain* 127. 2 (2004), pp. 243–58.

Boivin, Michael J., 'Finding God in Prozac or Finding Prozac in God: Preserving a Christian View of the Person amidst a Biopsychological Revolution', *Christian Scholar's Review* 32.2 (2003), pp. 159–76.

Borgmann, Albert, *Power Failure: Christianity in the Culture of Technology* (Grand Rapids, MI: Brazos, 2003).

—, *Technology and the Character of Everyday Life* (Chicago: University of Chicago Press, 1984).

Bostrom, Nick 'Astronomical Waste: The Opportunity Cost of Delayed Technological Development', *Utilitas* 15 (2003), pp. 308–14.

—, 'Human Genetic Enhancements: A Transhumanist Perspective', *The Journal of Value Inquiry* 37.4 (2003), pp. 493–506.

—, 'The Transhumanist FAQ: A General Introduction, Version 2.1' (2003), 55pp, http://transhumanism.org/index.php/WTA/faq/ (accessed 9 April, 2005), 4.

—, 'Transhumanist Values', www.nickbostrom.com/ethics/values.html.

Braun, Bruce, 'Querying Posthumanisms', *Geoforum* 35 (2004), pp. 269–72.

Breck, J., *The Sacred Gift of Life: Orthodox Christianity and Bioethics* (Crestwood, NY: St Vladimir's Seminary Press, 1998).

Breithaupt, Holger, and Caroline Hadley, 'Curing Ageing and the Consequences' (interview with Aubrey de Grey), *EMBO Reports* 6.3 (2005), pp. 198–201.

Brooke, J. Hedley and Geoffrey Cantor, *Reconstructing Nature* (Edinburgh: T & T Clark, 1998).

—, *Science and Religion: Some Historical Perspectives* (Cambridge: Cambridge University Press, 1991).

Brown, N., and M. Michael, 'An Analysis of Changing Expectations: or "Retrospecting Prospects and Prospecting Retrospects"', *Technology Analysis and Strategic Management* 15 (2003), pp. 3–18.

—, 'From Authority to Authenticity: The Changing Governance of Biotechnology', *Health, Risk and Society* 4 (2002), pp. 259–72.

Brown, N., B. Rappert and A. Webster, 'Introducing Contested Futures: From Looking into the Future to Looking at the Future', in N. Brown, B. Rappert and A. Webster (eds.), *Contested Futures: A Sociology of Techno-science* (Aldershot: Ashgate Press, 2000), pp. 3–20.

Buchanan, Alan, *et al.*, *From Chance to Choice: Genetics and Justice* (Cambridge: Cambridge University Press, 2000).

Butler, C., *Postmodernism* (Oxford: Oxford University Press, 2002).

Bynum, C.W., *The Resurrection of the Body in Western Christianity 200–1336* (New York: Columbia University Press, 1995).

Callahan, Daniel, 'Death and the Research Imperative', *New England Journal of Medicine* 34 (2000), pp. 654–56.

—, *False Hopes: Why America's Quest for Perfect Health is a Recipe for Failure* (New York: Simon & Schuster, 1998).

—, 'Aging and the Life Cycle: A Moral Norm?', in Callahan, ter Meulen and Topinkova (eds.), *A World Growing Old*, pp. 20–27.

Callahan, Daniel, Ruud H.J. ter Meulen and Eva Topinkova (eds.), *A World Growing Old: The Coming Health Care Challenges* (Washington, DC: Georgetown University Press, 1995).

Campbell, A., 'What is Wrong with Cloning Humans?' *Journal of Health Services Research Policy* 8.3 (July 2003), pp. 191–92.

Campbell, H., and M. Walker, 'Religion and Transhumanism: Introducing a Conversation', *Journal of Evolution and Technology* 14.2 (August 2005), pp. 1–14.

Canseco, J., *Juiced* (San Francisco: HarperCollins, 2005).

Caplan, Arthur, 'Open your Mind', *The Economist* (23 May 2002), pp. 73–75.

—, 'No Brainer: Can We Cope with the Ethical Ramifications of New Knowledge of the Human Brain?', in *Neuroethics: Mapping the Field*, Dana Foundation Conference Proceedings, San Francisco, 13–14 May 2002 (New York: Dana Press, 2002).

Caspi, A., *et al.*, 'Influence of Life Stress on Depression: Moderation by a Polymorphism in the 5-HTT Gene', *Science* 301 (2003), pp. 386–89.

de Caussade, J.P., *The Sacrament of the Present Moment*, trans. Kitty Muggeridge (London: Collins, 1981).

Changeux, J.P., and P. Ricoeur, *What Makes Us Think?* Trans M.B. DeBevoise (Princeton, NJ: Princeton University Press, 2000).

Chapman, A.R., *Unprecedented Choices* (Minneapolis: Fortress Press, 1999).

Chrysostom, J., *Homily on Romans*, at www.newadvent.org/fathers/210224.htm.

Clarkson, E.D., 'Fetal Tissue Transplantation for Patients with Parkinson's Disease: A Database of Published Clinical Results', *Drugs & Aging* 18 (2001), pp. 773–85.

Cole-Turner, R., *The New Genesis: Theology and the Genetic Revolution* (Louisville, KY: Westminster/John Knox Press, 1993).

Communion and Stewardship: Human Persons Created in the Image of God (The Vatican, International Theological Commission, Congregation for the Doctrine of the Faith, 2005), www.vatican.va/roman_curia/congregations/cfaith/cti_documents/rc_con_cfaith_doc.

Condit, C., *The Meanings of the Gene* (Wisconsin: University of Wisconsin Press, 1999).

Congregation for the Doctrine of the Faith, *Instruction on Respect for Human Life in Its Origin and on the Dignity of Procreation: Replies to Certain Questions of the Day* (Vatican: Holy See, 1987).

Conrad, P., 'Use of Expertise: Sources, Quotes and Voice in the Reporting of Genetics in the News', *Public Understanding of Science* 8 (1999), pp. 285–302.

Conradie, Ernst, *An Ecological Christian Anthropology: At Home on Earth?* (Basingstoke: Ashgate, 2005).

Cooper, David, 'Technology: Liberation or Ensnarement?', in R. Fellows (ed.), *Philosophy and Technology* (Cambridge: Cambridge University Press, 1995).

Crick, Francis, *The Astonishing Hypothesis: the Scientific Search for the Soul* (London: Simon & Schuster, 1994).

Cruz, Eduardo, 'The Nature of Being Human', in T. Peters and G. Bennett (eds.), *Bridging Science and Religion* (London: SCM Press, 2002), pp. 173–84.

Cryonics Institute, *Comparing Procedures and Policies.* www.cryonics.org/comparisons.html (accessed 19 June 2005).

Daly, T.T., 'Therapy vs. Enhancement: The Problem Posed by Anti-Aging Technologies', 2005, The Center for Bioethics and Culture Network, www.thecbc.org/redesigned/research_display.php?id=199.

Damasio, Antonio, *Looking for Spinoza: Joy, Sorrow and the Feeling Brain* (London: William Heineman, 2003).

—, *Descartes' Error: Emotion, Reason and the Human Brain* (New York: Putnam Berkley Group, 1994; London: Macmillan General Books, 1996).

Daniels, Norman, *Just Health Care* (New York: Cambridge University Press, 1985).

Darwin, C., *The Origin of Species* (London: John Murray, 6th edn, 1872).

Daston, Lorraine, and Katherine Park, *Wonders and the Order of Nature* (New York: Zone Books, 1998).

Davidoff, F., *et al.*, 'Evidence Based Medicine: A New Journal to Help Doctors Identify the Information They Need', *British Medical Journal* 310 (1995), pp. 1085–86.

Davis, Erik, *Techgnosis: Myth, Magic and Mysticism in the Age of Information* (London: Serpent's Tail, 1998).

Dawkins, R., *The Selfish Gene* (Oxford: Oxford University Press, rev. edn, 1989 [1975]).

Deane-Drummond, *Genetics and Christian Ethics* (Cambridge: Cambridge University Press, 2006).

—, *The Ethics of Nature* (Oxford; Blackwell, 2004).

—, (ed.), *Brave New World? Theology, Ethics, and the Human Genome* (London and New York: T & T Clark International, 2003).

—, *Biology and Theology Today* (London: SCM Press, 2001).

—, *Creation Through Wisdom: Theology and the New Biology* (Edinburgh: T & T Clark, 2000).

—, and Bronislaw Szerszynski, *Re-Ordering Nature: Theology, Society and the New Genetics* (London: T & T Clark, 2003).

Definition of Health, Preamble to the Constitution of the World Health Organization, adopted by the International Health Conference, New York, 19–22 June, 1946. www.who.int/about/definition/en/ (accessed 1 October 2005).

Dennett, Daniel C., *Darwin's Dangerous Idea: Evolution and the Meanings of Life* (London: Penguin, 1996).

Descartes, Rene, *The Philosophical Works of Descartes*, I (New York: Cambridge University Press, 1970).

'Disease Insights from Stem Cells' (Editorial), *Nature* 422, (24 April 2003), p. 787.

Dödsorsaker 2000. Socialstyrelsen: Sveriges officiella statistik, Hälsa och sjukdomar 2002:4. www.sos.se/FULLTEXT/42/2002-42-4/sammanfattning.htm (accessed 1 October 2005).

Donley, Carol C., 'Primary Literary Sources on Prolongevity', in Post and Binstock (eds.), *The Fountain of Youth*, pp. 433–43.

Dulap, W., 'Two Fragments: Theological Transformations of Law, Technological Transformations of Nature', in Carl Mitcham and Jim Grote (eds.), *Theology and Technology: Essays in Christian Analysis and Exegesis* (New York and London: Rowman & Littlefield, 1984).

Dulles, Avery, *Models of the Church* (New York: Doubleday, 2nd expanded edn, 1987).

Dupré, L., *Passage to Modernity* (London and New Haven: Yale University Press, 1995).

Dyson, A.O., 'At Heaven's Command? The Churches, Theology and Experiments on Embryos', in A.O. Dyson and J. Harris (eds.), *Experiments on Embryos* (London: Routledge, 1990), pp. 89–102.

Eccles, J.C., *The Human Psyche* (Berlin: Springer Verlag, 1980), pp. 27–49.

—, *The Human Mystery* (Berlin: Springer Verlag, 1978), pp. 214–29.

Eisenberg, Leon, 'Would Cloned Humans Really be Like Sheep?' *New England Journal of Medicine* 340.6 (1999), pp. 471–75.

Elliott, Carl, 'Pharma Goes to the Laundry: Public Relations and the Business of Medical Education', *Hastings Center Report* 34.5 (2004), pp. 18–23.

Ellul, Jacques, *The Technological Society*, trans. J. Wilkinson, intro R. Merton (London: Jonathan Cape, 1965).

Englehardt, H.T., *The Foundations of Christian Bioethics* (Liss: Swets & Zeitlinger, 2000).

Ettinger, Robert C.W., *The Prospect of Immortality.* www.cryonics.org/book1. html (accessed 19 June 2005).

European Molecular Biology Organization (EMBO), 'Interview: Curing Ageing and the Consequences', *EMBO Reports* 6.3 (2005), pp. 198–201.

Ezzell, Carol, 'Why? The Neuroscience of Suicide', *Scientific American* (February 2003), pp. 44–51.

Farah, M.J., *et al.*, 'Neurocognitive Enhancement: What Can We Do and What Should We Do?' *Nature Reviews. Neuroscience* 5 (2004), pp. 421–25.

Farquhar, D., *The Other Machine: Discourse and Reproductive Technologies* (New York: Routledge, 1996).

Feenberg, Andrew, *Critical Theory of Technology* (Oxford: Oxford University Press, 1991).

Fletcher, J., 'Indicators of Humanhood: A Tentative Profile of Man', *Hastings Center Report* 2 (September–October 1972), pp. 1–4.

Foerst, A., *God in the Machine: What Robots Teach Us About Humanity and God* (New York: Dutton, 2004).

Foster, Claire, 'Embryo Research: Some Anglican Perspectives', *Islam and Christian-Muslim Relations* 16 (2005), pp. 285–95.

Fukuyama, Francis, *Our Posthuman Future: Consequences of the Biotechnology Revolution* (New York: Farrar, Strauss and Giroux, 2002).

Funkenstein, A., *Theology and the Scientific Imagination in the Seventeenth Century* (Princeton, NJ: Princeton University Press, 1986).

Giles, Jim, 'Change of Mind', *Nature* 430 (2004), p. 14.

Goering, S., 'Gene Therapies and the Pursuit of a Better Human', *Cambridge Quarterly of Health Care Ethics* 9 (2000), pp. 330–41.

Goodfield, J., *Playing God: Genetic Engineering and the Manipulation of Life* (New York: Harper & Row, 1977).

Graham, Elaine L., *Representations of the Post/Human: Monsters, Aliens and Others in Popular Culture* (Manchester: Manchester University Press, 2002).

Graham, G., *Genes: A Philosophical Inquiry* (London and New York: Routledge, 2002).

—, *Ethics and International Relations* (Oxford: Blackwell, 1997).

Gregory of Nyssa, *The Life of Moses*, trans. A.J. Malherbe and E. Ferguson (New York: Paulist, 1978).

de Grey, Aubrey D.N.J., 'An Engineer's Approach to Developing Real Anti-Aging Medicine', in Post and Binstock (eds.), *The Fountain of Youth*, pp. 249–67.

Grove, Peter, ' "Vereinigungsphilosophie" beim frühen Schleiermacher und bei Herder', in Ulrich Barth and Klaus-Dieter Osthövener (eds.), *200 Jahre, Reden über die Religion* (Berlin and New York: W. de Gruyter, 2000), pp. 328–43.

Gunton, Colin E., *The Triune Creator: A Historical and Systematic Study* (Grand Rapids, MI: Eerdmans, 1998).

Habermas, Jürgen, *The Future of Human Nature* (Cambridge: Polity Press, 2003).

Hacking, Ian, *Representing and Intervening: Introductory Topics in the Philosophy of Science* (Cambridge: Cambridge University Press, 1983).

Hagell, P., and P. Brundin, 'Cell Survival and Clinical Outcome Following

Intrastriatal Transplantation in Parkinson Disease', *Journal of Neuropathology and Experimental Neurology* 60 (2001), pp. 741–52.

Halberstam, J., and I. Livingston (eds.), *Posthuman Bodies* (Indianopolis: Indiana University Press, 1995).

Haldane, J., and P. Lee, 'Aquinas on Human Ensoulment, Abortion and the Value of Life', *Philosophy* 78 (2003), pp. 255–78.

Hall, Stephen S., *Merchants of Immortality: Chasing the Dream of Human Life Extension* (Boston, MA and New York: Houghton Mifflin Co., 2002).

Hamilton, M. (ed.), *The New Genetics and the Future of Man* (Grand Rapids, MI: Eerdmans, 1972).

Haraway, Donna J., *Simians, Cyborgs, and Women: The Reinvention of Nature* (New York: Routledge, 1991).

Harding, S., and J. O'Barr (eds.), *Sex and Scientific Inquiry* (Chicago: Chicago University Press, 1987).

Harris, John, 'Immortal Ethics', *Annals of New York Academy of Science*, 1019 (2004), pp. 527–34.

Hauerwas, Stanley, *Suffering Presence: Theological Reflections on Medicine, the Mentally Handicapped and the Church* (Edinburgh: T & T Clark, 1988).

—, *The Peaceable Kingdom* (London: SCM Press, 1984).

Haught, John F., *Deeper than Darwin* (Boulder, CO: Westview Press, 2003).

Hayles, N. Katherine, *How We Became Posthuman: Virtual Bodies in Cybernetics, Literature, and Informatics* (Chicago: University of Chicago Press, 1999).

Hedgecoe, A., 'Transforming Genes: Metaphors of Information and Language in Modern Genetics', *Science as Culture* 8 (1999), pp. 209–29.

Hefner, Philip, *Technology and Human Becoming* (Minneapolis: Fortress, 2003).

Heidegger, Martin 'The Question of Technology' [1954], in D.F. Krell (ed.), *Basic Writings* (London: Routledge, 1993).

—, *Being and Time*, trans. John Macquarrie and Edward Robinson (Oxford: Blackwell, 1978 [1962]). German original: *Sein und Zeit* (Halle: Niemeyer, 1927).

Henderson, L., and J. Kitzinger, 'The Human Drama of Genes: "Hard" and "Soft" Representations of Breast Cancer Genetics', *Sociology of Health and Illness* 21.5 (1999), pp. 560–78.

Herder, Johann Gottfried, *Essay on the Origin of Language*, in J.J. Rousseau and J.G. Herder, *On the Origin of Language*, trans. J.H. Moran and A. Gode (New York: Bergman, 1967).

—, *Outlines of a Philosophy of the History of Man*, trans. T. Churchill (London, 1800; Chicago: University of Chicago Press, 1966).

Herissone-Kelly, P., 'The Cloning Debate in the United Kingdom: The Academy Meets the Public', *Cambridge Quarterly of Healthcare Ethics* 14 (2005), pp. 268–79.

Herzfeld, N.L., *In Our Image: Artificial Intelligence and the Human Spirit* (Minneapolis: Fortress Press, 2002).

Hook, C.C., 'The Techno Sapiens are Coming', *Christianity Today* 48.1 (January 2004), pp. 38–39.

Hope, Tony, 'Evidence Based Medicine and Ethics', *Journal of Medical Ethics* 21.5 (1995), pp. 259–60.

Hrdy, Sarah Blaffer, *Mother Nature: A History of Mothers, Infants and Natural Selection* (New York: Pantheon, 1999).

Hudson, J., 'What Kinds of People Should We Create?', *Journal of Applied Philosophy* 17 (2000), pp. 131–43.

Hume, D., *A Treatise of Human Nature* (London: Penguin Classics, 1985).

Huxley, A., *Brave New World* (London: Chatto and Windus, 1932).

Iverson, J., *In Search of the Dead: A Scientific Investigation of Evidence for Life after Death* (London: BBC Books, 1992).

Jackelén, Antje, *Time and Eternity* (Philadelphia: Templeton Foundation Press, 2005).

—, 'The Image of God as *Techno Sapiens*', *Zygon* 37.2 (2002), pp. 289–302.

Jeeves, Malcolm (ed.), *From Cells to Souls – and Beyond: Changing Portraits of Human Nature* (Grand Rapids, MI and Cambridge: Eerdmans, 2004).

Jones, D. Gareth, *Designers of the Future: Who Should Make the Decisions?* (Oxford: Monarch Books, 2005).

—, 'The Emergence of Persons', in Malcolm Jeeves (ed.), *From Cells to Souls – and Beyond* (Grand Rapids, MI: Eerdmans, 2004), pp. 11–33.

Jones, D. Gareth, and Kerry Galvin, 'Neural Grafting in Parkinson's Disease: Scientific and Ethical Ambiguity', in preparation.

Jones, D. Gareth, and Sharon Sagee, 'Xenotransplantation: Hope or Delusion?' *Biologist* 48 (2001), pp. 129–32.

Jones, David A., *The Soul of the Embryo* (London and New York: Continuum, 2004).

Juengst, E.T., 'What Does *Enhancement* Mean?', in Erik Parens (ed.), *Enhancing Human Traits: Ethical and Social Implications* (Washington, DC: Georgetown University Press, 1998).

Jung, Carl Gustav, *The Archetypes and the Collective Unconscious*, Collected Works, 9 (London: Routledge & Kegan Paul, 1959).

—, *Psychological Approach to the Dogma of the Trinity*, Collected Works, 11 (London: Routledge & Kegan Paul, 1958).

Junker-Kenny, Maureen, 'Genetic Enhancement as Care or as Domination? The Ethics of Asymmetrical Relationships in the Upbringing of Children', *Journal of Philosophy of Education* 39 (2005), pp. 1–12.

—, 'Die "Anschauung des Universums...zur Vollkommenheit ausgebildet". Zur Christologie der "Reden"', in Ulrich Barth and Klaus-Dieter Osthövener (eds.), *200 Jahre, Reden über die Religion* (Berlin and New York: W. de Gruyter, 2000), pp. 714–28.

—, 'The Transcendental Turn: Shifts in Argumentation between the First and the Second Editions of the Glaubenslehre', in *New Athenaeum/Neues Athenaeum* (Lewiston, NY: Edwin Mellen Press, 1992), pp. 21–41.

—, *Das Urbild des Gottesbewusstseins. Zur Entwicklung der Religionstheorie und Christologie Schleiermachers von der ersten zur zweiten Auflage der Glaubenslehre* (Berlin and New York: W. de Gruyter, 1990).

Kass, Leon (ed.), *Being Human: Core Readings in the Humanities* (New York: W.W. Norton, 2004).

Kass, Leon R., *Life, Liberty and the Defense of Dignity: The Challenge for Bioethics* (San Francisco, CA: Encounter Books, 2002).

Kelsey, David, 'Human Being', in P. Hodgson and Robert King (eds.), *Christian Theology: An Introduction to its Traditions and Tasks* (London: SPCK, 1988).

Kirkwood, T.B.L., *Time of Our Lives: The Science of Human Ageing* (Oxford: Oxford University Press, 1999).

Kitzinger, J., 'Media Templates: Patterns of Association and the (Re)construction of Meaning over Time', *Media, Culture and Society* 22 (2000), pp. 64–84.

Kitzinger, J., L. Henderson, A. Smart and J. Eldridge, 'Media Coverage of the Social and Ethical Implications of Human Genetic Research'. Final report for The Wellcome Trust, 2003.

Koch, T., 'Disability and Difference: Balancing Social and Physical Constructions', *Journal of Medical Ethics* 27 (2001), pp. 370–76.

Kramer, Peter D., *Listening to Prozac* (New York: Penguin Books, 1994).

Kurzweil, Ray, *Singularity is Near: When Humans Transcend Biology* (New York: Viking, 2005).

—, *The Age of Spiritual Machines: When Computers Exceed Human Intelligence* (New York and London: Penguin Books, 2000).

—, *The Age of Spiritual Machines: When Computers Exceed Human Intelligence* (New York: Viking, 1999).

Latour, B., *We Have Never Been Modern* (London: Prentice Hall, 1993).

—, *We Have Never Been Modern*, trans. C. Porter (Cambridge, MA: Harvard University Press, 1993).

—, *Science in Action: How to Follow Scientists and Engineers Through Society* (Cambridge, MA: Harvard University Press, 1987).

LaVan, David A., Terry McGuire and Robert Langer, 'Small Scale Systems for In Vivo Drug Delivery', *Nature Biotechnology* 21 (2003), pp. 1184–91.

Lemmens, Trudo, 'Piercing the Veil of Corporate Secrecy about Clinical Trials', *Hastings Center Report* 34.5 (2004), pp. 14–18.

Leplin, Jarrett, 'Realism and Instrumentalism', in W.H. Newton-Smith (ed.), *A Companion to the Philosophy of Science* (Oxford: Blackwell, 2000), pp. 393–401.

Levitt, P., B. Reinoso and L. Jones, 'The Critical Impact of Early Cellular Environment on Neuronal Development', *Preventive Medicine* 27 (1998), pp. 180–83.

Lewis, J., 'What Counts in Cultural Studies?', *Media, Culture and Society* 19 (1997), pp. 83–97.

Lewis, J., and T. Speers, 'Misleading Media Reporting?: The MMR Story', *Nature Reviews Immunology* 3 (2003), pp. 913–18.

Lippert-Rasmussen, K., 'Measuring the Disvalue of Inequality Through Time', *Theoria* 69 (2003), pp. 32–45.

Louth, Andrew, *The Origins of the Christian Mystical Tradition: From Plato to Denys* (Oxford: Oxford University Press, 1981).

Löwendahl, Lena, *Med kroppen som instrument* (Stockholm: Almqvist & Wiksell International, 2002).

Lucke, Jayne C., and Wayne Hall, 'Who Wants to Live Forever?', *EMBO Reports* 6.2 (2005), pp. 98–102.

Luria, Alexander R., *The Man with a Shattered World*, trans. L. Solotaroff; (Harmondsworth: Penguin Books, 1975).

Macdonald, F., and C.H.J. Ford, *Oncogenes and Tumour Suppressor Genes* (Oxford: Bias Scientific Publishers, 1991).

MacIntyre, Alasdair, *Whose Justice? Which Rationality?* (London: Duckworth, 1988).

—, *After Virtue: A Study in Moral Theory* (London: Duckworth, 2nd edn, 1985).

Manson, N., 'The Precautionary Principle, the Catastrophe Argument and Pascal's Wager', *Ends and Means* 4.1 (1999).

Mantzaridis, Gergios, *The Deification of Man* (New York: St Vladimir's Seminary Press, 1984).

Martin, Brian, 'Scientific Fraud and the Power Structure of Science', *Prometheus* 10.1 (1992), pp. 83–98, www.uow.edu.au/arts/sts/bmartin/pubs/92prom.html (accessed 5 August 2005).

Martinson, Brian C., Melissa S. Anderson and Raymond de Vries, 'Scientists Behaving Badly', *Nature* 435 (2005), pp. 737–38.

May, William F., 'The Medical Covenant: An Ethics of Obligation or Virtue?', in Gerald P. McKenny and Jonathan R. Sande (eds.), *Theological Analyses of the Clinical Encounter* (Dordrecht: Kluwer, 1994), pp. 29–44.

Mayberg, H.S., 'Modulating Dysfunctional Limbic-cortical Circuits in Depression: Towards Development of Brain-based Algorithms for Diagnosis and Optimised Treatment', *British Medical Bulletin* 65 (2003), pp. 193–207.

McKenny, Gerald P., *To Relieve the Human Condition: Bioethics, Technology and the Body* (Albany, NY: State University of New York Press, 1997).

Mehta, M.A., *et al.*, 'Methylphenidate Enhances Working Memory by Modulating Discrete Frontal and Parietal Lobe Regions in the Human Brain', *Journal of Neuroscience* 20 (2000), RC65: 1–6.

Merleau-Ponty, M., *Phenomenology of Perception* (London: Routledge & Kegan Paul, 1962; French original, 1945).

Messer, Neil G., 'Healthcare Resource Allocation and the "Recovery of Virtue"', *Studies in Christian Ethics* 18.1 (2005), pp. 89–108.

—, 'Human Genetics and the Image of the Triune God', *Science and Christian Belief* 13.2 (2001), pp. 99–111.

Milbank, John, 'Knowledge: The Theological Critique of Philosophy in Hamann and Jacobi', in John Milbank, Catherine Pickstock and Graham Ward (eds.), *Radical Orthodoxy: A New Theology* (London: Routledge, 1999), pp. 21–37.

Miller, D., J. Kitzinger and K. Williams, *The Circuit of Mass Communication: Media Strategies, Representation and Audience Reception in the AIDS Crisis* (London: Sage, 1998).

Miller, Franklin G., and Howard Brody, 'A Critique of Clinical Equipoise: Therapeutic Misconception in the Ethics of Clinical Trials', *Hastings Center Report* 33.3 (2003), pp. 19–28.

Mission and Public Affairs Council of the Church of England, *Embryo Research: Some Christian Perspectives* (London: Archbishop's Council, 2003).

Mitchel, C.B., 'Define "Better"', *Christianity Today* 48.1 (January 2004), pp. 42–45.

—, *Science and Wisdom* (London: SCM Press, 2003).

Moltmann, J., *In the End – the Beginning* (London: SCM Press, 2004).

—, 'The World in God, or God in the World?', in R. Bauckham (ed.), *God Will be All in All: An Eschatology of Jürgen Moltmann* (Edinburgh: T & T Clark, 1999).

—, *The Coming of God* (London: SCM Press, 1996).

—, *The Way of Jesus Christ: Christology in Messianic Dimensions* (London: SCM Press, 1990).

—, *Creating a Just Future* (London: SCM Press, 1989).

—, *Theology of Hope* (London: SCM Press, 1967).

Moravec, Hans, *Robot: Mere Machine to Transcendent Mind* (Oxford and New York: Oxford University Press, 1999).

—, *Mind Children: The Future of Robot and Human Intelligence* (Cambridge, MA and London: Harvard University Press, 1988).

Murdoch, A. 'Off the Treadmill – Leaving an IVF Programme Behind', in Jocelynne A. Scutt (ed.), *Baby Machine: Reproductive Technology and the Commercialisation of Motherhood* (London: Green Print, 1990), pp. 66–73.

Murray, Stephen B., 'Reimagining Humanity: The Transforming Influence of Augmenting Technologies Upon Doctrines of Humanity', in M. Breen, E. Conway and B. McMillan (eds.), *Technology and Transcendence* (Dublin: The Columba Press, 2003), pp. 195–216.

Nisbet, M., D. Brossard and A. Kroepsch, 'Framing Science: The Stem Cell Controversy in an Age of Press/Politics', *The Harvard International Journal of Press/Politics* 8 (2003), pp. 36–70.

Nordenfelt, Lennart, *On the Nature of Health: An Action-theoretic Approach* (Dordrecht: Reidel, 2nd edn, 1995).

Northcott, M., 'Concept Art, Clones and Co-Creators: The Theology of Making', *Modern Theology* 21 (2005), pp. 219–36.

Novaes, Simone Batement and Tania Salem, 'Embedding the Embryo', in John Harris and Søren Holm (eds.), *The Future of Human Reproduction: Ethics, Choice and Regulation* (Oxford: Oxford University Press, 1998), pp. 100–26.

O'Donovan, Oliver, *Begotten or Made?* (Oxford: Oxford University Press, 1984).

—, *Resurrection and Moral Order: An Outline for Evangelical Ethics* (Leicester: InterVarsity Press, 1986).

O'Mathuna, D.P., 'Genetic Technology, Enhancement, and Christian Values', *The National Catholic Bioethics Quarterly* 2.2 (Summer 2002), pp. 277–95.

Olshansky, S. Jay, and Bruce A. Carnes, 'In Search of the Holy Grail of Senescence', in Post and Binstock (eds.), *The Fountain of Youth*, pp. 133–59.

—, *The Quest for Immortality: Science at the Frontiers of Aging* (New York and London: Norton, 2001).

Palmer, Michael, *Freud and Jung on Religion* (London and New York: Routledge, 1997).

Pannenberg, Wolfhart, *Anthropology in Theological Perspective*, trans. M. O'Connell (Philadelphia, PA: Westminster, 1985).

Parfit, D., *Reasons and Persons* (Oxford: Oxford University Press, 1984).

Parnia, S., D. Waller, R. Yeates and P. Fenwick, 'A Qualitative Study of the Incidence, Features and Aetiology of Near Death Experiences in Cardiac Arrest Survivors', *Resuscitation* (2001), pp. 149–55.

Penfield, W., *The Mystery of the Mind* (Princeton, NJ: Princeton University Press, 1975).

Peters, T. *Playing God? Genetic Determinism and Human Freedom* (London and New York: Routledge, 2nd edn, 2002).

—, *GOD – The World's Future: Systematic Theology for a New Era* (Minneapolis: Fortress Press, 2nd edn, 2000).

—, (ed.), *Genetics: Issues of Social Justice* (Cleveland: Pilgrim Press, 1997).

—, *Playing God?: Genetic Determinism and Human Freedom* (London and New York: Routledge, 1997).

Peters, T., R.J. Russell and M. Welker (eds.), *Resurrection: Theological and Scientific Assessments* (Grand Rapids, MI: Eerdmans, 2002).

Petersen, A., 'Replicating our Bodies, Losing our Selves: News Media Portrayals of Human Cloning in the Wake of Dolly', *Body and Society* 8 (2002), pp. 71–90.

Peterson, G.R., *Minding God: Theology and the Cognitive Sciences* (Minneapolis: Fortress Press, 2003).

Peterson, J.C., *Genetic Turning Points: The Ethics of Human Genetic Intervention* (Grand Rapids, MI: Eerdmans, 2001).

Pezawas, L., *et al.*, '5-HTTLPR Polymorphism Impacts Human Cingulate-amygdala Interactions: A Genetic Susceptibility Mechanism for Depression', *Nature Neuroscience* 8 (2005), pp. 828–34.

Philipson, Sten M., and Nils Uddenberg (eds.), *Hälsa som livsmening* (Stockholm: Natur och kultur, 1989).

Philo, G. (ed.), *Message Received* (Harlow: Addison, Wesley and Longman, 1999).

Polkinghorne, J., and M. Welker (eds.), *The End of the World and the Ends of God* (Harrisburg, PA: Trinity Press International, 2000).

Pope John Paul II, *Evangelium Vitae* (Vatican: Holy See, 1995).

—, 'The Ethics of Genetic Manipulation', *Origins* 13.23 (1983), pp. 386–89.

Porter, M., 'Shadow and Light: Reflections on Life with Psychotropic Medications', *Sojourners* (March–April 1998), p. 25.

Post, Stephen G., and Robert H. Binstock (eds.), *The Fountain of Youth: Cultural, Scientific, and Ethical Perspectives on a Biomedical Goal* (New York: Oxford University Press, 2004).

Ramsey, P., 'Genetic Therapy: A Theologian's Response', in Michael Hamilton (ed.), *The New Genetics and the Future of Man* (Grans Rapids, MI: Eerdmans, 1972).

—, *Fabricated Man: The Ethics of Genetic Control* (New Haven: Yale University Press, 1970).

Reich, Jens, *Es wird ein Mensch gemacht. Möglichkeiten und Grenzen der Gentechnik* (Berlin: Rowohlt, 2003).

Rheingold, H., *Virtual Reality* (New York: Simon and Schuster, 1991).

Richardson, Ruth D., *The Role of Women in the Life and Thought of the Early Schleiermacher (1768–1806)* (Lewiston: Edwin Mellen Press, 1991).

Rifkin, J., *The Biotech Century: How Genetic Commerce Will Change the World* (London: Phoenix, 1999).

Robertson, John A., *Children of Choice: Freedom and the New Reproductive Technologies* (Princeton, NJ: Princeton University Press, 1994).

Rohls, J., *Geschichte der Ethik* (Tubingen: Mohr-Siebeck, 1991).

Rose, Michael R., 'Laboratory Evolution of Postponed Senescence in Drosophila Melanogaster', *Evolution* 38 (1984), pp. 1004–10.

Rose, Michael R., Hardip B. Passananti and Margarida Matos (eds.), *Methuselah Flies: A Case Study in the Evolution of Aging* (New Jersey: World Scientific, 2004).

Ross, J.E., *et al.*, 'The Effect of Genetic Differences and Ovarian Failure: Intact Cognitive Function in Adult Women with Premature Ovarian Failure versus Turner Syndrome', *Journal of Clinical Endocrinology and Metabolism* 89.4 (2004), pp. 1817–22.

Rossouw, J.E., G.L. Anderson, R.L. Prentice, *et al.*, 'Risks and Benefits of Estrogen Plus Progestin in Healthy Postmenopausal Women: Principal Results from the Women's Health Initiative Randomized Controlled Trial', *Journal of the American Medical Association* 288.3 (2002), pp. 321–33.

Rothman, S.M. and D.J. Rothman, *The Pursuit of Perfection: The Problems and Perils of Medical Enhancement* (New York: Pantheon Books, 2004).

Sacks, Oliver, *An Anthropologist on Mars* (New York: Arnold A. Knopf, 1995).

—, *The Man Who Mistook His Wife for a Hat* (London: Gerald Duckworth, 1985).

Sandberg, A., *Morphological Freedom – Why We not just Want it, but **Need** it* (2001), www.nada.kth.se/~asa/Texts/MorphologicalFreedom.htm (accessed 27 September 2005).

Schaefer, Susanne, *Gottes Sein zur Welt* (Regensburg: Pustet, 2002).

Scharff, Robert C., and Val Dusek (eds.), *Philosophy of Technology: An Anthology* (Oxford: Blackwell, 2003).

Scheper-Hughes, N., 'Commodity Fetishism in Organs Trafficking', in N. Scheper-Hughes and L. Wacquant (eds.), *Commodifying Bodies* (London: Sage, 2002), pp. 31–62.

Schleiermacher, Friedrich Daniel Ernst, 'Outline for a Reasonable Catechism for Noble Women' (1798), trans. R.D. Richardson, in Ruth D. Richardson, *The Role of Women in the Life and Thought of the Early Schleiermacher (1768–1806)* (Lewiston: Edwin Mellen Press, 1991), pp. 60–61.

—, *On Religion: Speeches to its Cultured Despisers*, intro., trans. and notes Richard Crouter (Cambridge: Cambridge University Press, 1988).

—, *The Christian Faith*, ET of 2nd German edn 1830–31, ed. H.R. Mackintosh and J.S. Stewart (Edinburgh: T & T Clark, 1986).

Schumacher, J.M., *et al.*, 'Transplantation of Embryonic Porcine Mesencephalic Tissue in Patients with Parkinson's Disease', *Neurology* 54 (2000), pp. 1042–50.

Schwöbel, C., and C.E. Gunton (eds.), *Persons, Divine and Human* (Edinburgh: T & T Clark, 1991).

Scott, Peter, *A Political Theology of Nature* (Cambridge: Cambridge University Press, 2003).

Sherrington, C.S., *Man on his Nature* (Cambridge: Cambridge University Press, 1940).

Shostak, Stanley, *Becoming Immortal: Combining Cloning and Stem-cell Therapy* (Albany, NY: SUNY Press, 2002).

—, *Becoming Immortal: Combining Cloning and Stem-Cell Therapy* (Albany: State University of New York Press, 2002).

Siep, Ludwig, 'Genomanalyse, menschliches Selbstverständnis und Ethik', in Ludger Honnefelder and Peter Propping (eds.), *Was wissen wir, wenn wir das menschliche Genom kennen?* (Köln: Dumont, 2001), pp. 196–205.

Silver, L.M., *Remaking Eden: How Genetic Engineering and Cloning Will Transform the American Family* (New York: Avon, 1998).

Simkin, Benjamin, *Medical and Musical Byways of Mozartiana* (Santa Barbara, CA: Fithian Press, 2004).

Simon, Bart, 'Toward a Critique of Posthuman Futures', *Cultural Citique* 53 (2003), pp. 1–9.

Skuse, D., K. Elgar and E. Morris, 'Quality of Life in Turner Syndrome is Related to Chromosomal Constitution: Implications for Genetic Counselling and Management', *Acta Paediatrica Supplement* 88 (4280) (1999), pp. 110–13.

Skuse, D.H., *et al.*, 'Evidence from Turner's Syndrome of an X-linked Locus Affecting Cognitive Function', *Nature* 387 (6634) (1997), pp. 705–708.

Smart, A., 'Reporting the Dawn of the Post-genomic Era: Who Wants to Live Forever?', *Sociology of Health and Illness* 25 (2003), pp. 24–49.

Smith, D.H., 'Creation, Preservation and all the Blessings...', *Anglican Theological Review* 81 (1999), pp. 567–88.

Smyth, Nicholas, *Charles Taylor: Meaning, Morals, and Modernity* (Cambridge: Polity Press, 2002).

Sperry, R.W., *Neurosciences Third Study Program* (Cambridge and London: MIT Press, 1974).

Spinoza, Benedictus de, *The Ethics Part II* (New York: Dover Press, 1955).

'Staff Background Paper: Human Genetic Enhancement', US President's Council on Bioethics (December 2002), www.bioethics.gov/.

Statistiska Centralbyrån, Sweden, *Befolkningsstatistik, Återstående medellivslängd för åren 1751–2004.* www.scb.se/templates/tableOrChart_____25830.asp (accessed 1 October 2005).

Stojkovic, M., *et al.*, 'Derivation, Growth and Applications of Human Embryonic Stem Cells', *Reproduction* 128 (2004), pp. 259–67.

Strathern, M., 'The Meaning of Assisted Kinship', in M. Stacey (ed.), *Changing Human Reproduction: Social Science Perspectives* (London: Sage, 1992), pp. 148–69.

Sumner, L. Wayne, *Welfare, Happiness, and Ethics* (Oxford: Clarendon Press, 1996).

Swift, Jonathan, *Gulliver's Travels, into several remote nations of the world* (London, 1726).

Takala, C., 'The (Im)Morality of (Un)Naturalness', *Cambridge Quarterly of Healthcare Ethics* 13 (2004), pp. 15–19.

Taylor, Charles, *Sources of the Self: The Making of Modern Identity* (Cambridge: Cambridge University Press, 1989).

Teilhard de Chardin, Pierre, *Christianity and Evolution* (New York: Harcourt Brace and Co., 1968).

Temkin, L., *Inequality* (Oxford: Oxford University Press, 1993).

Tertullian, *Apology*, trans. S. Thelwall, *Anti-Nicene Fathers*, Vol III, at www.ccel.org.fathers2/ANF-03/anf03-05.htm p.253_53158.

—, *The Soul*, trans. Peter Holmes, *Anti-Nicene Fathers*, Vol III, at www.ccel.org/fathers2/ANF-03/anf03-22.htm.

Timm, Hermann, *Gott und die Freiheit. Studien zur Religionsphilosophie der Goethezeit. Bd. I. Die Spinozarenaissance* (Frankfurt: Klostermann, 1974).

Tipler, F.J., *The Physics of Immortality* (New York: Doubleday, 1994).

Torrance, Thomas F., *The Being and Nature of the Unborn Child* (Lenoir: Glen Lorien Books, 2000).

United States Office of Science and Technology Policy, 'Federal Policy on Research Misconduct', www.ostp.gov/html/001207_3.html (accessed 5 August 2005).

Urs von Balthasar, H., *Theo-Drama: Theological Dramatic Theory. V. The Final Act*, trans. Graham Harrison (San Francisco: Ignatius Press, 1998).

Van Lommel P., R. Van Wees, V. Meyers and I. Elfferich, 'Near-death Experience in Survivors of Cardiac Arrest: A Prospective Study in the Netherlands', *Lancet* 358 (2001), pp. 2039–2045.

Vardy, Peter, *Being Human: Fulfilling Genetic and Spiritual Potential* (London: Darton, Longman & Todd, 2003).

Verhey, Allen, *Reading the Bible in the Strange World of Medicine* (Grand Rapids, MI: Eerdmans, 2003).

Vogel, G., 'Stem Cell Research: Cloning Pioneer Heads Toward Human Frontier', *Science* 298 (4 October 2002), pp. 37–39.

Voltaire, *Candide, or, Optimism*, trans. and ed. Theo Cuffe (London: Penguin, 2005).

Walford, Roy L., *Maximum Life Span* (New York and London: W.W. Norton, 1983).

Walter, J.J., 'Catholic Reflections on the Human Genome', *The National Catholic Bioethics Quarterly* 3.2 (Summer 2003), pp. 275–86.

Walters, L., and J.G. Palmer, *The Ethics of Human Gene Therapy* (Oxford and New York: Oxford University Press, 1997).

Waters, Brent, *From Human to Posthuman: Christian Theology and Technology in a Postmodern World* (London: Ashgate, 2006).

—, 'From *Imago Dei* to Technosapien? The Theological Challenge of Transhumanism', unpublished paper, 2004.

Wells, S., *Transforming Fate Into Destiny: The Theological Ethics of Stanley Hauerwas* (Carlisle: Paternoster Press, 1998).

Wenz, Peter, 'Engineering Genetic Injustice', *Bioethics* 19.1 (2005), pp. 1–11, www.blackwell-synergy.com/loi/biot.

Wesley, J., *A Plain Account of Christian Perfection* (London: Epworth, 1952).

Wiese, Christian, '"Dass man zusammen Philosoph und Jude ist..." Zur Dimension des Jüdischen in Hans Jonas' philosophischer Ethik der Bewahrung der "Schöpfung"', in Joachim Valentin and Saskia Wendel (eds.), *Jüdische Traditionen in der Philosophie des 20. Jahrhunderts* (Darmstadt: Wiss. Buchgesellschaft, 2000), pp. 131–47.

Wilkinson, S., 'Commodification Arguments for the Legal Prohibition of Organ Sale', *Health Care Analysis* 8 (2000), pp. 189–201.

Williams, Anna N., *The Ground of Union: Deification in Aquinas and Palamas* (Oxford: Oxford University Press, 1999).

Williams, C., J. Kitzinger and L. Henderson, 'Envisaging the Embryo in Stem Cell Research: Discursive Strategies and Media Reporting of the Ethical Debates', *Sociology of Health and Illness* 25 (2003), pp. 793–814.

Williams, Rowan, 'Macrina's Deathbed Revisited: Gregory of Nyssa on Mind and Passion', in Lionel Wickham and Caroline P. Bammel (eds.), *Christian Faith and Greek Philosophy in Ancient Antiquity* (Leiden: E.J. Brill, 1993), pp. 227–46.

Wilmut, I., *et al.*, 'Viable Offspring Derived from Fetal and Adult Mammalian Cells', *Nature* 385 (27 February 1997), pp. 810–813.

Wilson, E.O., *On Human Nature* (London: Penguin Books, 1995).

—, *Sociobiology: The New Synthesis* (Cambridge, MA: Harvard University Press, 1975).

Wolpe, Paul Root, '*Neurotechnology*, Cyborgs, and the Sense of Self', in *Neuroethics: Mapping the Field*, Dana Foundation Conference Proceedings, San Francisco, 13–14 May 2002 (New York: Dana Press, 2002).

World Council of Churches, *Manipulating Life: Ethical Issues in Genetic Engineering* (Geneva: World Council of Churches, 1982).

World Transhumanist Association, *Transhumanist FAQ 2.1* (2003), www.transhumanism.org/resources/FAQv21.pdf (accessed 27 Sept. 2005).

World Transhumanist Association. www.transhumanism.org (accessed 1 October 2005).

INDEX

Index notes: Some sub-sections are listed in page order to aid coherence of argument. Superscript numbers appended to page locators refer to notes' numbers.

Wenz, Peter, 180
Weslyan Methodism, 15
Williams, Anna, 177
Williams, Clare, xii, 130
Wilmut, Ian, 74
Wilson, E.O., 36

Wilson, Robert A., 171
wisdom, 16, 23, 178–9
  and foolishness, 39
Wolpe, Paul, 20
World Council of Churches, 20
World Health Organization (WHO), 144

Printed in Great Britain
by Amazon.co.uk, Ltd.,
Marston Gate.